Tentsmuir

Ten Thousand Years of Environmental History

Robert M. M. Crawford

Archaeopress Publishing Ltd
Summertown Pavilion
18-24 Middle Way
Summertown
Oxford OX2 7LG

www.archaeopress.com

ISBN 978-1-78969-124-5
ISBN 978-1-78969-125-2 (e-Pdf)

© Archaeopress and R.M.M. Crawford 2019

Cover: The Great Slack, photo courtesy of Tom Cunningham and Sea Eagle, photo courtesy of John Cummins. Leuchars Castle from a drawing in 1785 by William Crawford.

All rights reserved. No part of this book may be reproduced, or transmitted, in any form or by any means, electronic, mechanical, photocopying or otherwise, without the prior written permission of the copyright owners.

Printed in England by Severn, Gloucester

This book is available direct from Archaeopress or from our website www.archaeopress.com

To the Memory of

Dr John Berry
CBE, DL, FRSE, LLD, D.Sc, Ph.D

1907- 2002

First Director of the
Nature Conservancy in Scotland

and

creator in 1954 of the
Tentsmuir National Nature Reserve

At Tentsmuir in 1999

Contents

Preface		v
Acknowledgements		vi
Chapter One	Tentsmuir in prehistory	1
	Mesolithic Tentsmuir	2
	Mesolithic shell gatherers	4
	Norwegian Holocene Storegga Slide tsunami	5
	Neolithic, Bronze and Iron-age Tentsmuir	9
Chapter Two	Tentsmuir in history	12
	Medieval Tentsmuir	12
	The Normans at Leuchars	13
	Late medieval Tentsmuir	18
	Tentsmuir - a royal hunting forest	19
	Salmon fishing rights	19
	Agricultural improvement	22
	The draining of Tentsmuir	23
	Malaria at Tentsmuir	25
	Tentsmuir's farms	26
	German spies at Tentsmuir	29
Chapter Three	Sand and water	31
	Sand and water interactions	31
	Afforestation of Tentsmuir	33
	Dune slacks	35
	Permanent wetlands	36
	Longshore drift	37
	Rapid coastal accretion	40
	Coastal accretion time-scale	42

	Erosion	43
	Long-term history of the Tentsmuir coast	46
	The future survival of Tentsmuir	49

Chapter Four — Tentsmuir's dunes – a changing landscape — 50

	Dune structure and stability	50
	Physical versus biological fragility	53
	Sand dune grasses	54
	Plant geography at Tentsmuir	56
	Grey Dunes	62
	Woody plants on sand dunes	62
	Plant survival in sand dunes	66
	Drought tolerance	66
	Conclusions	69

Chapter Five — Tentsmuir's wetlands — 70

	Dune slacks and salt marshes	70
	Avoidance of anoxic stress	71
	Dune-slack variation	75
	Flood-line alder association	75
	Salt slacks	76
	Lotus corniculatus slack	79
	Erica tetralix slacks	80
	The desiccation of the slacks	80

Chapter Six — Land, people and resources — 84

	Tentsmuir as a sporting estate	84
	Tentsmuir ornithology	84
	The Tentsmuir Grouse Moor	87

Grouse Disease		90
Bird migration		93
20th century Tentsmuir		95
The afforestation of Tentsmuir		95
Tentsmuir drainage		98
Morton Lochs		101
Creation of Morton Lochs Nature Reserve		101
Tentsmuir forestry		104
Water table levels		105

Chapter Seven — Tentsmuir's thriving birds — 111

- Bird habitats at Tentsmuir — 111
- Winter visitors — 111
- Tentsmuir's geese — 113
- Shore birds — 117
- Birds of fresh waters — 120
- Birds of the forest — 123
- Raptors — 123
- Morton Lochs restored — 128
- Tentsmuir's ornithological future — 130

Chapter Eight — Tentsmuir's declining birds — 131

- Bird losses from Tentsmuir — 131
- Declining wader populations — 134
- Plovers — 137
- Wigeon — 138
- Historical bird records from the Eden Estuary — 138
- Wildfowl and lead poisoning — 139
- Physiology of swan-song — 139

Chapter Nine	Tentsmuir's mammals, butterflies and moths	141
	Seals	141
	Lepidoptera	146
	Rodentia	157
	Badgers	162
Chapter Ten	Saving the Wilderness	164
	Tentsmuir's origins	164
	Creating a Wilderness Reserve	165
	Ecological history of Tentsmuir Point	166
	Dune invasion by Scots Pine	168
	Biological birch control	168
	Onset of coastal erosion	169
	The Great Slack	171
	Erosion at the Eden Estuary	172
	Ecological restoration	175
	Conservation and education	178
	Environmental recording	179
	Tentsmuir in the future	180
References		183
Index		190

Preface

The Tentsmuir Peninsula lies between the estuaries of the rivers Tay and Eden and is remarkable for being one of the largest natural areas of wind-blown sand and dunes in Scotland. This peninsula stretches for just over 7 miles from north to south and almost 3.5 miles from east to west and is outstanding for being entirely a post ice-age development. The post-glacial advance of the coastline seawards has not been a continuous process. There have been both advances and retreats of the shoreline, which continue to take place.

Despite the fragile nature of this sandy coastal terrain, Tentsmuir has been a scene of human activity for over 10,000 years. It witnessed one of the earliest known occurrences in Scotland of Mesolithic hunter-gatherers and has supported human activities throughout the Neolithic, Bronze and Iron Ages. In medieval times it was in turn, a home for the Norman nobility, a royal hunting forest with highly valued fishing rights, as well as being a summer grazing for local pastoralists with tents (hence the early name *Tentis muris* (Taylor with Márkus, 2012).

In more modern times it was valued by the early agricultural improvers of the 18th century for the manner in which the land warmed in summer. A combination of high sunshine levels due to the shelter of the mountains to the north and west, and the sandy nature of the soil made it one of earliet areas in Scotland for harvesting grain in the 19th century.

Despite the activities of agricultural improvers, the blockage of drainage by the front line of dunes causes a significant rise in the level of the winter water table. Consequently, agriculture, and forestry are limited in the extent to which they can be profitably pursued. This restriction, together with the rigours of the coastal environment was important in preserving natural refugia for a wide range of plants, as well as resident and migrating birds and other animals. Tentsmuir's mid-Scotland location also offers habitats suitable for both northern and southern species which contributes greatly to the biodiversity of this outstanding setting for wildlife conservation. The uniqueness of this coastal region, therefore led to the creation of a National Nature Reserve of 92 acres at the north-eastern end of the Tentsmuir Peninsula in 1954. Since then, an active period of coastal accretion more than trebled the size of the Reserve, which is now unfortunately eroding in places. For the future, the probability of rising sea levels and increasing exposure to storms may cause a level of destruction such that the physical existence and biological future of Tentsmuir cannot be guaranteed.

Acknowledgements

A work of this wide range could not have been undertaken without the stimulation and encouragement provided by Scottish Natural Heritage and in particular Tom Cunningham, Manager of the Tentsmuir Reserve when this book was written. His help in searching out information, contacts, and illustrative material has been invaluable, He has also read every chapter with his wife Peta, to the book's great advantage. Numerous colleagues and friends have also read and helped with various chapters. I am greatly indebted to Professors John and Bryony Coles, Dr. Torben Bjarke Ballin and the late Alan Saville for the pre-historical section. I had generous advice on place names from early times to the present day from Dr. Simon Taylor. I am also greatly indebted to Professor Christopher Smout and my wife Barbara for historical advice.

Dr. Anne-Marie Smout and Dr. Ron Summers were of great assistance in relation to the rich bird-life of Tentsmuir. Professor Stephen Buckland was extremely generous in providing many fine bird photographs. An account of the butterflies and moths would not have been possible without the benefit of the detailed knowledge that was generously shared by David Bryant and Gillian Fyfe. Many past students from St Andrews University Botany Department assisted in mapping physical changes and measuring water table levels at Tentsmuir over the past 50 years.

I am particularly indebted to Mr. Willian Berry (6th of Tayfield) for allowing me access to historical documents and family photographs. The family of Tentsmuir's outstanding naturalist and artist, the late Len Fullerton, kindly gave permission to reproduce a selection of their father's evocative illustrations of Tentsmur's wildlife. The final version of the text was greatly improved with Dr. Bill Starkey's careful attention to detail for which I am very grateful.

Chapter One

Tentsmuir in prehistory

Where the estuaries of the rivers Tay and Eden flow into the sea there lies a peninsula, entirely made of sand and water, shaped over the millennia by constant confrontation with changing ocean levels and tidal currents. This large sandy region now called Tentsmuir is an ecological palimpsest in that it has a history that has been written, erased, and rewritten by constant interaction between sand and water. World-wide, the ice age covered the polar regions of the Earth in deep deposits of ice to such an extent that the sea level was lowered globally by up to 120m.

With the passing of the last Glacial Maximum c.18,000 years ago, the ice started to melt and sea levels gradually recovered to their pre-Pleistocene level. This global increase in sea-level which brought about a world-wide coastline retreat is referred to scientifically as eustatic rise. However, this was gradually reversed as the pressure of the ice on the Earth's crust lessened and the land began to rebound. This recovery of land from marine inundation is referred to geologically as isostatic rise. The extent of the Holocene isostatic land-rise has varied from place to place as during the Pleistocene period the depth of the ice would not have been even, and therefore the depression of the land would not have been uniform due to differences in the weight of the ice.

At Tentsmuir, as elsewhere, there would have been advances and retreats. Temporal and spatial patterns of relative sea-level change in the north of Britain and Ireland have occurred during four major episodes. Each episode was marked by a rise in sea level, causing a retreat of the shore-line, followed by a period of isostatic rebound, which usually more than compensated for the previous eustatic coastal retreat (Smith et al., 2012).

In Scotland many examples of rising land levels are seen in raised beaches of varying age and height. West

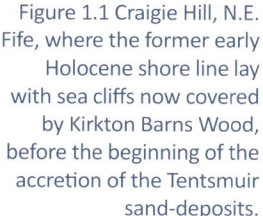

Figure 1.1 Craigie Hill, N.E. Fife, where the former early Holocene shore line lay with sea cliffs now covered by Kirkton Barns Wood, before the beginning of the accretion of the Tentsmuir sand-deposits.

of Tentsmuir, an early post-Pleistocene coastline location is visible as a former sea-cliff (Craigie Hill) on the west side of the B945 road to Tayport. This former Pleistocene-Holocene boundary sea-cliff is now 3.5 miles from the sea (Figure 1.1).

Mesolithic Tentsmuir

It is not clear just when the sea retreated from this ancient cliff at Craigie Hill. It is evident however that by 10,000 years ago the coastline had advanced a considerable distance seawards. It was here that a coastal Mesolithic presence was discovered on the high ground of the 'Old Quarry' field approximately 2.5 km northeast of the cliff at Craigie. This discovery came to light when some flints were first collected from molehills and other exposures. This site of early human activity at Tentsmuir is now just under 2.5 miles inland from the sea. However, at the time of its early human occupation it was adjacent to the shore, either as a peninsula or possibly an island (Figures 1.2-3).

This early site (Morton A) was first excavated between 1963-7 by the original discoverer R.D.M. Candow with Dundee Museum and Art Gallery. The site was then further investigated in 1969-70 by J.M. Coles and included on this occasion an additional trench (Morton B). Further excavation of these two sites (Morton A and B) have revealed what is apparently a place of early human Mesolithic seasonal camping activity (Candow, 1989).

Further excavations at both sites found numerous artefacts, including various small and large end-scrapers, burins (engravers), awls and a variety of microliths (Coles, 1971). Hammerstones, together with grinding and polishing stones were found and also hearths, and stake holes, which probably served as the fastenings for wind-breaks around sleeping areas. It would appear that at the time of this early occupation, the high ground of the 'Old Quarry' field was a low island, 225 by 75 m linked to the mainland only at low tide. Whether or or not hunting groups from differing localities shared the site together at any one time is not clear.

The first of the two sites that were investigated (Morton A) was near the highest part of the former promontory at 12 m (OD) while the second (Morton B) lay on the northern slope of the promontory, 40 m NE of Morton A. It would appear from the bivouac nature of the sleeping arrangements, that these sites were probaby occupied mainly in summer for

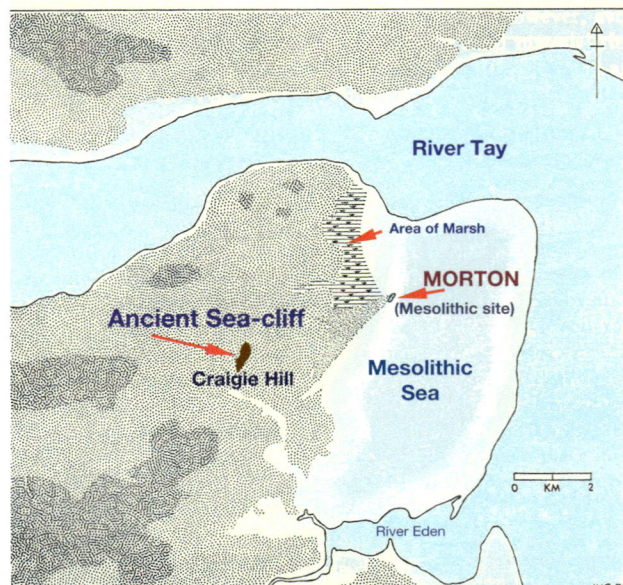

Figure 1.2 Conjectural map of the Morton area, c. the 5th millennium BC showing Morton as an island. Stippled areas are land over 50 feet O.D., rising inland to over 300 ft. (double stipple). To the northwest of Morton is an area of marsh. Blue shading indicates the present position of the sea. The light partially coloured area would most probably have been the sea at the time of the last Mesolithic human occupation. (Map reproduced and adapted with permission from Coles et al., 1971).

Figure 1.3 Site of the excavation of the Mesolithic site at Morton.

accessing food resources from this maritime location. The stone tools of Morton contain raw materials from St Monans, Wormit, the Sidlaws and Ochil Hills and the southern Tay shore and Fife coast as well as stone of local coastal or stream gravel origin (Coles, 1983).

The promontory covers an area of 6000 m², of which probably under 400 m² were occupied, and of this only 60 m² were stratigraphically excavated at site A sufficiently to detect different occupation horizons (Coles, 1983). The typology of the flint microliths is diagnostic of the Early Mesolithic and suggests that this first phase of the Mesolithic occupation of Morton is most likely to predate *circa* 8500 cal. BC as after this time there was a change to a later type of lithic tool (Saville, 2008). The early microliths were in the shape of isosceles triangles (*in which it is the two shorter sides that are of equal length*), but after 8500 cal. BC they were replaced by a more elongated form of triangle (*in which the two longer sides are the ones of equal length*).

These Mesolithic sites would have been cold and exposed in winter. It is probable however, from the species composition of the mollusc population which co-habited the site with its human population, that the climate would have been relatively dry with fallen timbers and deep shade. Similar settlements along the north shore of the River Tay might represent temporary occupations, either from Morton or other contemporary people (Coles, 1983).

The excavations at Morton (Figure 1.4) have revealed evidence also of extensive marine beach deposits which indicate that the location of the site at the time of occupation was close to the main post-glacial shoreline (Coles, 1971). Fish bones included; Cod, Haddock, Turbot, Sturgeon, Salmon (or Sea Trout). The bones of birds were mostly of species that inhabit open water which included Fulmar, Gannet, Cormorant, Shag, Puffin, Razorbill, and Guillemot. Thrush and Crow (either carrion or hooded) were also found. It was suggested that some of these species may have been caught at sea or else found after being driven ashore by storms. The animal remains included red and roe deer, wild cattle and wild boar, as well as hedgehogs and bank voles.

Some evidence was found for the use of plants, which, in addition to Hazel nuts included, Orache (*Atriplex patula*), Fat Hen (*Chenopodium album*) and Corn Spurrey (*Spergula arvensis*). The physical evidence in relation to the nature of the site suggests that it would most likely have been used for a series of seasonal occupations of the promontory over periods lasting between 200-500 years during an overall 2000 year history of periodic visitations. The small number of stone artefacts found on each occupation surface was taken to indicate the temporary nature of the camps (see Figure 1.5).

The number of people occupying the site at any one time and collaborating in building fences and having animal-drives or even constructing weirs for fish traps, could not be accurately estimated but was thought to be in the region of 40-50 persons (Coles, 1983).

Mesolithic shell gatherers

Morton B contained a large quantity of seashells. Similar large shell midden mounds have been found around the coasts of Britain and Ireland. However, it is on the shores of Mainland Scotland and the Western Isles that have some of the largest deposits indicating the importance of shellfish in the Mesolithic diet. These have yielded further information as to the seasonal activities of the Mesolithic shell gatherers as the shells of the common cockle grow at varying rates at different times of the year, depending on air and sea temperatures and the salinity of the water. Given a constant supply of food experimentally, the greatest growth rates of cockles have been found to be in spring and early summer (Ibarrola et al., 2008).

A study of these shells at Morton has shown that a high

Figure 1.4 Excavations in progress at Morton (1969) with the Cambridge supervisor and three ladies from Dundee. (Photo courtesy of John Coles.)

Figure 1.5 A Dundee Museum artist's impression of Mesolithic Morton (Reproduced with permission from Candow (1989).

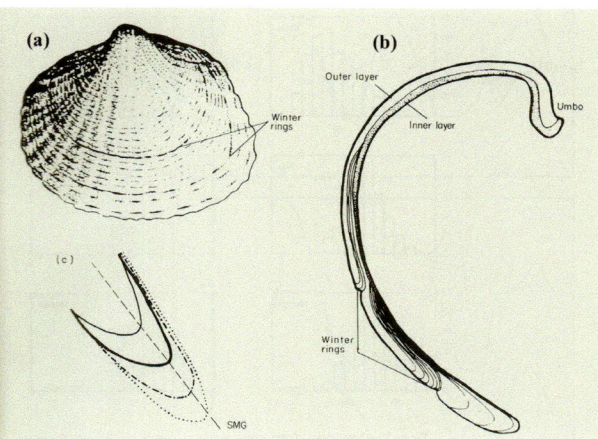

Figure 1.6 Shells of the Common Cockle (*Cerastoderma edule*). (a) Shell showing two winter rings on the outer surface of the shell. (b) A radial section shows the winter rings as indentations in the shell. (c) The growing edge in section, shows the last two increments laid down (continuous lines) and the two about to be deposited (dotted lines). (Reproduced with permission from Deith 1986).

proportion of them had been harvested high on the beach and although no particular season was found, there was evidence within the structure of the shells (Figure 1.6) that the cockles were harvested in winter as well as summer even although those harvested in winter would have been less nutritious than those harvested in summer (Deith, 1983, 1986).

An additional feature of Mesolithic shell fish harvesting is the fact that the shells were not just dumped, but were used to create large mounds. It has been suggested that these almost monumental-like middens may have had a symbolic purpose as they could have served for marking out the position of a particular band's stretch of beach (Deith, 1986). Given that Morton was not a permanent residential site, but one that was visited regularly over a long period, would indicate that already by the Mesolithic the very human attribute of having a sense of property, and the need to identify such ownership had already become part of human behaviour.

These excavations (Morton A and B) have revealed two major phases of Mesolithic occupation. The earlier phase was abandoned and covered by wind-blown sand before a later phase. Samples of bevel-ended bone tools from the Morton B site have provided radiocarbon (C-14) dates which point to the 5th millenium BC for this later occurrence and suggest that there could have been a gap in human occupation at Morton of 4000 years (Bonsall et al., 1995; Saville, 2004). The length of this gap is strange, in that it suggests that no people visited Morton for such a long time, even although the region was readily accessible (see below).

Mid-Mesolithic remains in Fife are not unique to Morton. Not far away at Fife Ness, evidence of Mesolithic activity has been found with C-14 dates from around the middle of the 8th millennium BC (Wickham-Jones and Dalland, 1998). Other Mesolithic sites have been found at Broughty Ferry, on the other side of the Tay Estuary from Tentsmuir and also at Aberdour near South Queensferry. This latter site was discovered during the excavations for the second Forth Road Bridge Queensferry Crossing. Here, remnants of a Mesolithic hut circle were found on the south bank of the Forth. The dwelling, probably for seasonal use, as at Morton, was based around an oval pit approximately 7 metres (23 ft) in length, which has been dated to around 8240 BC, making it the earliest known dwelling in Scotland (Cowing, 2012).

Norwegian Holocene Storegga Slide tsunami

Since these excavations were carried out there has been an accumulation of evidence for the deposition of massive sand and silt deposits at many sites on the east coast of Scotland. At first, some of these deposits were thought to have been the result of an *'exceptional tide.'* It is now realized that these deposits were caused by a widespread mid-Holocene coastal-flood caused by a tsunami generated by a submarine slide on the continental slope off the mid-western coast of Norway (Figures 1.7-8) referred to as the *Norwegian Holocene Storegga Slide tsunami*.

This struck the east coast of Scotland with a 70 feet (21 m) high tsunami along a stretch of coastline over 600 km long (Smith et al., 2010).

Subsequent percolation of peat deposits into the Scottish tsunami layers have caused some confusion in defining an exact date for the tsunami. However, recent studies based on the Norwegian deposits indicate a probable date of 8,100-8200 cal. BP (Vasskog et al., 2013).

Tsunamite, (the deposit left by a *tsunami*) dating from this event can be found at various Scottish coastal locations, including some Shetland Islands and is today a feature in the Montrose Basin, where there is a layer of deposited sand about 0.6 metres (2.0 ft) thick. (Figures 1.7-8).

The *Norwegian Holocene Storegga Slide tsunami* left its mark on Tentsmuir. Towards the western edge of the Tentsmuir Peninsula near Craigie, the rising ground consists of raised beaches from the last glacial period (Late Devensian). These raised beaches, are dissected by a number of gullies leading eastwards which are now filled with peat bogs. The largest of these is Silver Moss which lies within St Michaels Wood. Here a series of varied deposits of geological detritus have been found with prominent layers of grey, micaceous, fine

Figure 1.7 Deposit 2 feet thick in Montrose Basin caused by the Storegga tsunami (Photo courtesy of Wikipedia Commons).

sand. As with other similar deposits, they were at first thought to be due to storm surges, but have now been identified as being a result of the *Norwegian Holocene Storegga Slide tsunami*.

At, or about the time of the last *Storegga Slide*, there was an area of land known to archaeologists as 'Doggerland.' This name was given by B.J. Coles (Coles, 1988) for the land that once linked Great Britain, Denmark and the Netherlands across what is now the southern North Sea (Figure 1.8). The Dogger Bank is so named from the Midde Dutch name for a Cod Fishing boat. The *Doggerland* is believed to have included within its coastline areas of lagoons, marshes, mudflats, and beaches, which provided rich hunting, fowling and fishing grounds for Mesolithic peoples (Coles, 1998; Coles, 2000). Although much of Doggerland was already physically submerged by the time of the tsunami due to a gradual rise in sea level (Ballin, 2016) it has been suggested that the

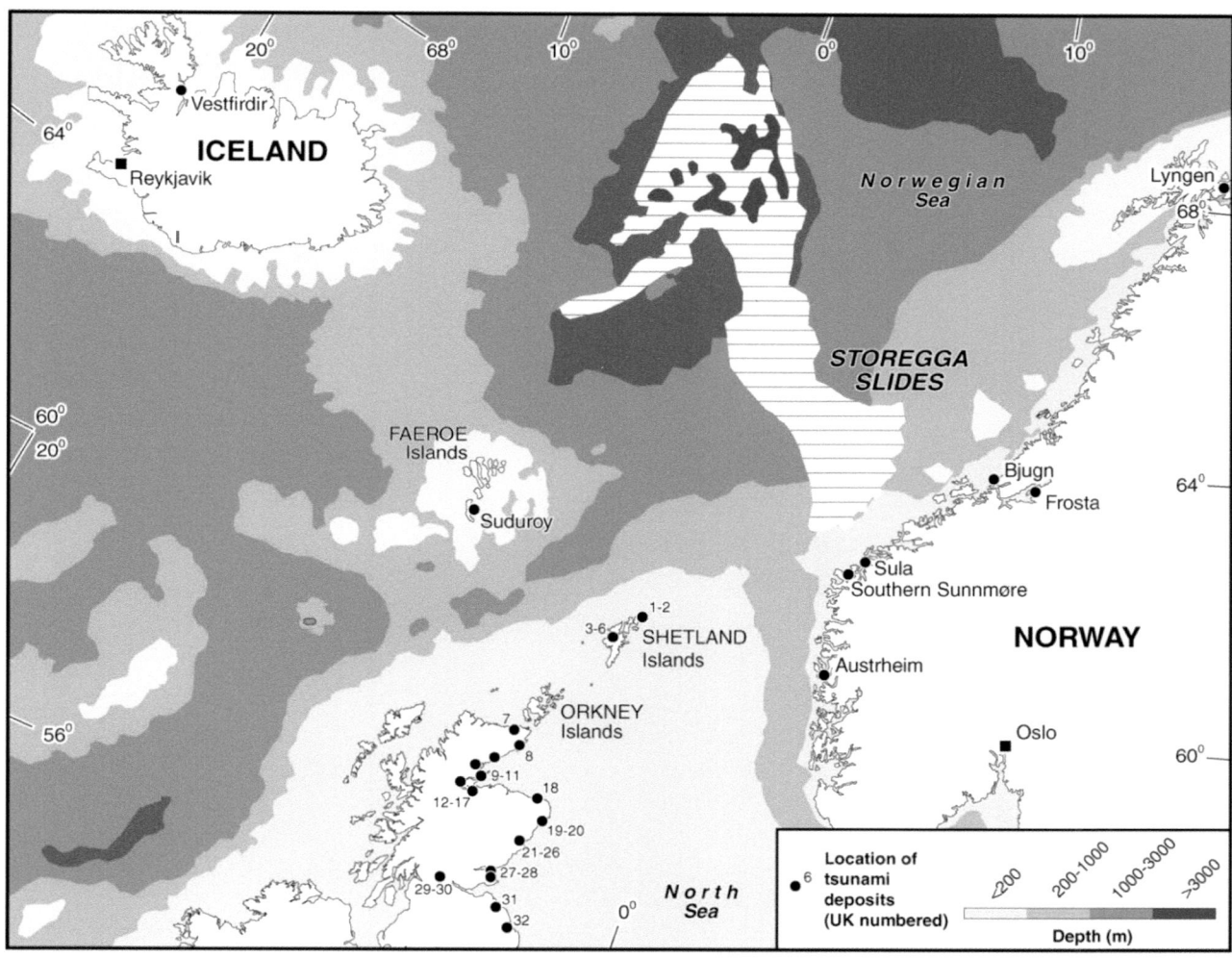

Figure 1.8 Location of the Storegga Slides and sites where evidence for the Holocene Storegga Slide tsunami has been found in the United Kingdom (reproduced with permission from Smith et al., 2012).

tsunami would have had a catastrophic impact on a significant part of the contemporary Mesolithic populations. Widespread destruction and drowning would have taken place in coastal areas of both Britain and mainland Europe (Figure 1.9). There would also have been severe restrictions on the movements of the human Mesolithic populations in what are now the Low Countries of the European continent. This landmass, the Doggerland, which is now covered by the North Sea, has been thought to have had an important influence on the course of prehistory in northwestern Europe.

Opinions differ in relation to the gradual inundation of Doggerland by rising sea levels. Professor Bryony Coles, who invented the term Doggerland, produced a hypothetical map (Figure 1.10) of the major estuaries and the coastline that would have been attractive for Mesolithic human settlements (Coles, 1998). Various attempts at more precise outlines showing the progressive retreat of the Doggerland coastline have subsequently appeared (Ballin and Bjerck, 2016) but as yet there is no unanimity or precision in this matter. The development of a maritime-based society along the northern coast of Doggerland, would have had specialized adaptations to this zone, including an ability to move with changing sea-levels. It has been suggested that such a mobile culture might have delayed the spread of farming into coastal regions (Coles, 2000).

The question remains as to from where the Scottish Mesolithic people originated. Recent studies have compared with great precision the form of the late Palaeolithic and early Mesolithic small stone tools (microliths) usually made of flint or chert and have been able to define a difference in style from those found in northern Britain from these found in the south. The differences in these artefacts strongly suggest that northern Britain was settled or visited from the north-east, by groups belonging to the Scandinavian Palaeolithic-Early Mesolithic complex.

There are different views as to the development of Doggerland between c. 16,000-8,000 BC (see Figures 1.9-10; Ballin and Bjerck, 2016; Coles, 1998). However, following the map presented in Figure 1.9 it would have been possible for people to walk almost the entire way from Norway to Scotland in a straight line.

Archaeological excavations carried out at Morton, on the former peninsula just to landward of the mid-Holocene shoreline, disclosed two phases of Mesolithic occupation. The earlier phase, as dated from the typology of the flint microliths is now thought to predate c. 8500 cal. BC and would therefore have

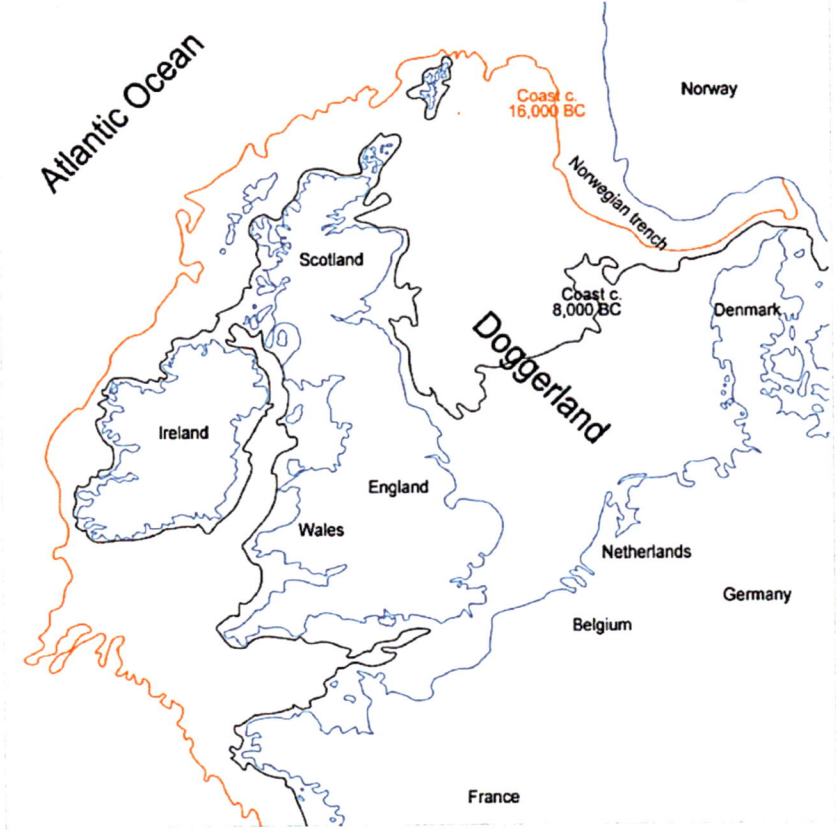

Figure 1.9 Map of Doggerland, showing the staged retreat of the Doggerland coastlines at ca. 16,000 BC and ca. 8000 BC (according to Spinney et al 2012). The red lines indicate the coast around 16,000 BC, and the black lines around 8000 BC. (Reproduced with permission from Ballin and Bjerck, 2016).

Tentsmuir in prehistory

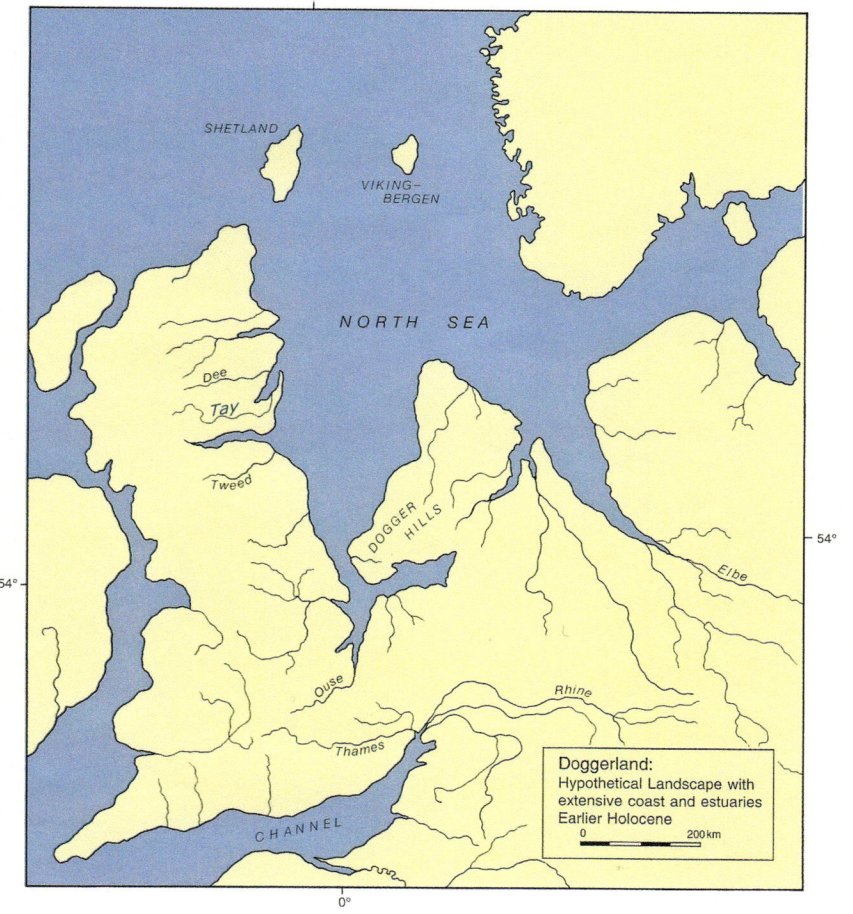

Figure 1.10 A hypothetical map of Doggerland and surrounding areas of northwest Europe in the Earlier Holocene. The major estuaries and the coastline would have been attractive for Mesolithic human settlements (Coles, 1988). Map devised by B.J. Coles & S.E. Rouillard. Copyright B.J. Coles and S.E. Rouillard.

catastrophe of the *Storegga Slide tsunami* would have isolated the Mesolithic inhabitants in the North Sea. Whether or not this brought about an avoidance of coastal sites by early peoples, causing the long-lasting absence of Mesolithic activity for several millennia as noted at Morton, is a matter of speculation.

The Morton Mesolithic site is now nearly 4 km from the sea, which indicates that there has been an advance seawards of the shore-line at Tentsmuir which has averaged 0.5 m per annum over a period of *c.* 8000 years. Such a rate of accretion would appear quite possible when put into context with recent coastal changes. At the end of the 20th century the shoreline at the north-eastern region of Tenstsmuir advanced seawards by several metres per annum (see Chapter 2).

Neolithic, Bronze and Iron Age Tentsmuir

Tentsmuir continued to be a site for human activity throughout prehistory. Evidence for this is found largely in the western part of the peninsula (Figure 1.11) where numerous finds and crop marks have been found (Carter, 1997).

Already in 1905 it had been noted that whenever the soil is disturbed large fragments of coarse pottery were turned up regularly over the whole location (see Longworth et al., 1966). Such finds have included a cinerary urn, and numerous beakers. In the 19th century the remains of a fine corded beaker were discovered on the Earlshall Estate (Figure 1.12).

The numerous items found include, Neolithic pottery, grooved ware, and beakers specifically for domestic use. The stone implements comprise leaf-shaped flint

been a pre-tsunami activity which was abandoned and covered by wind-blown sand before a later post-tsunami occupational phase.

The circumstances of abandonment of the earlier phase are unclear, although it was noted that the site had become more exposed to the sea. If the tsunami was involved, the dates from the excavations here would place it between the two occupation phases of Mesolithic Morton. The question therefore arises,, did the Main Holocene Transgression play a role and make occupation impossible for a period? If so, the

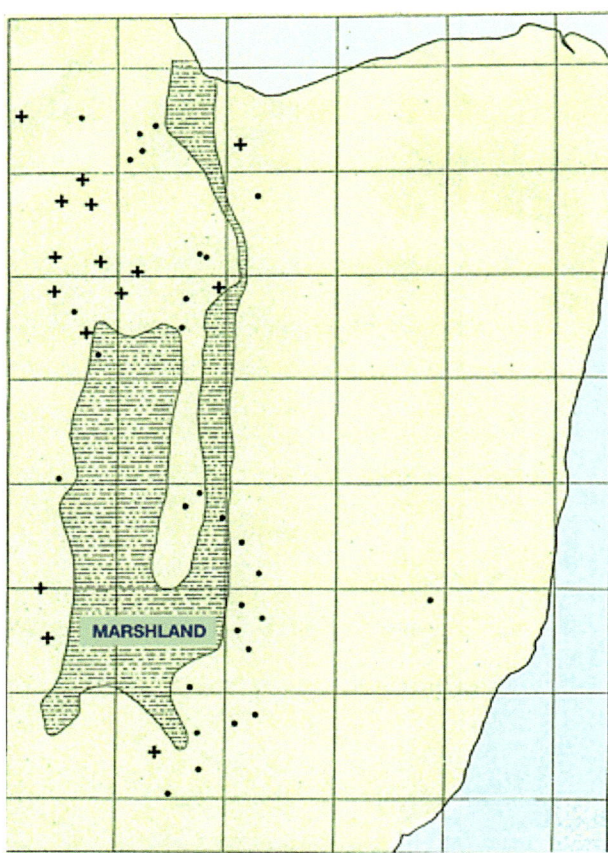

+ Cropmark of prehistoric type
• Artefact

marks and artefacts in the eastern parts of present-day Tentsmuir (Figure 1.11) would indicate that the prehistoric coastline had not at this time advanced to this region. There is only one record of a small prehistoric deposit found in this eastern region by Candow (Carter, 1997).

The implements found include, barbed-and-tanged flint arrowheads with numerous bronze artefacts including many pins. Other finds comprise, beads of vitreous paste, jet, amber, and glass, all found singly, and ornaments of jet, including armlets. A possible mid-Bronze Age blade was found in association with an urn (Coles, 1964). Shell mounds have frequently been noticed but it is not always certain that they are of human origin. Excavations of some minor shell middens has suggested that they were merely a product of collapsed and eroded sand-hills (Shand, 1908).

A record of early dwellings in the shape of hut circles was reported in the 18th century in the 1st

Figure 1.11 Distribution at Tentsmuir of Neolithic, Bronze Age, and Iron Age sites and artefacts (Reproduced from Headland Archaeology Ltd unpublished report (1999).

arrowheads and polished stone axes. Bronze Age sherds have been found at Shanwell and on Garpit Links (Longworth et al., 1966-7).

Neolithic finds at Tentsmuir comprise pottery, including beakers, food vessels, collared and cordoned urns, as well as large food vessels. The absence of crop

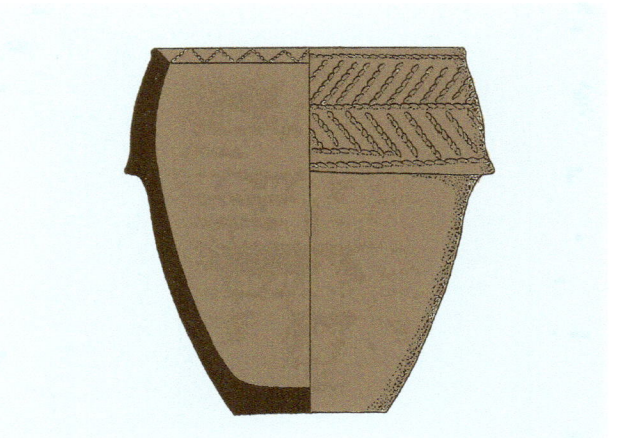

Figure 1.12 Reconstruction of a corded beaker found on Earlshall Estate made of coarse paste tampered with grit and with a smooth light brown surface. Decoration: on the collar, twisted cord lines. In the rim, twisted cord zigzag. (Reproduced with permission from Longworth et al., 1966-7).

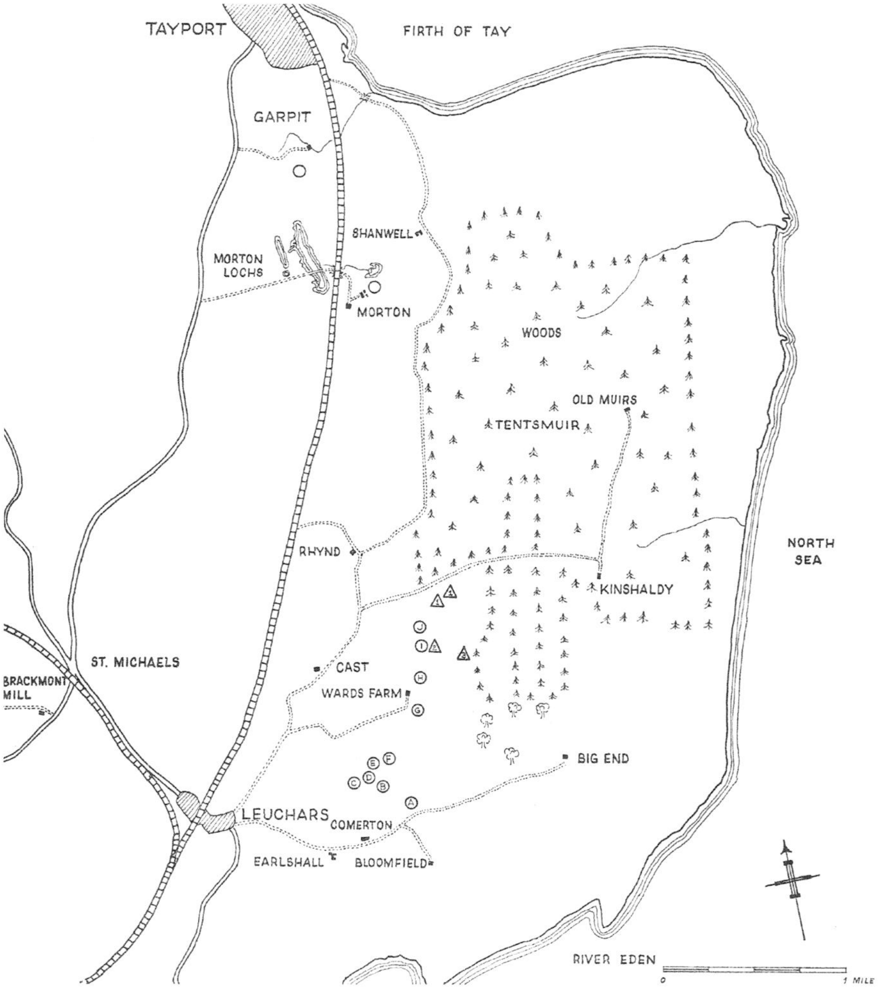

Figure 1.13 Map of Tentsmuir including locations of Neolithic, Bronze and Iron Age artefacts mentioned in text. (Reproduced with permission from Longworth et al., 1966).

Statistical Account by the Reverend Kettle the Minister of Leuchars (Kettle, 1796). These were described as circular hollows 10-15 feet in diameter, surrounded by earthen or turf mounds and have been subsequently interpreted as prehistoric round-houses, but are no longer visible. They may have been genuine prehistoric structures that have unfortunately been destroyed by a wave of agricultural improvement at Tentsmuir in the latter half of the 18th century and forestry plantations in the 20th century. No traces of these structures have survived as confirmation of such dwellings.

The nature and quantity of these finds has led to the conclusion that Tentsmuir was settled perhaps intermittently by differing peoples from Late Neolithic times and through much of the ensuing Bronze Age. The majority of the finds come from sand bunkers north and south of Wards Farm and the Earlshall Estate, as well as from south of Morton Lochs and Garpit (see Fig. 1.13).

Chapter Two

Tentsmuir in history

Medieval Tentsmuir

The extent of human occupation of Tentsmuir becomes clearer by the medieval period (11th century onwards). As with most prehistorical sites, our knowledge of their existence is based on excavations together with fragments of pottery and tools that have been found, as well as other more dateable objects. At Tentsmuir these range from items such as a coin of Edward the Confessor (1041-66) found at Kinshaldy (Talyor with Márkus, 2010, pp 520-1) and other coins dating from Henry III of England (1248) to Charles II (1660-85) and later. More substantial survivals include buildings and historical documents.

Tentsmuir first enters into written history in the mid XII century when between 1150 and 1152 King David I makes a grant of Fetters (*Fotheros- Fetterdale*) *besides St Andrews with its proper boundaries* to Dunfermline Abbey. Shortly afterwards between 1182 and 1187, Leuchars is recorded when the Church there along with various lands is granted to the Priory of St Andrews by Ness son of William on the understanding that he would be buried at St Andrews (Taylor et al., 1984) pp. 513

The name of Tentsmuir itself first appears in use in the 14th and 15th centuries, One example is *'Tyntis Mure'* which is recorded in 1455-65 when the Barclays of *Tyntis Mure* in 1437 fled to France after the murder of James I. They they were then subsequently executed by the Duke of Brittany (Talyor with Márkus, 2010, pp. 547-8).

Place-names (toponomy), when correctly understood and associated with documents, yield information of past human activities that lie embedded within their meaning and derivation. Tentsmuir lies in the Parish of Leuchars which is notable as it has several of the older Scots place-names that are recorded in the earliest records of the late 12th century (Taylor and Márkus, 2010).

To attempt to interpret place-names, without carefully considering their earliest forms, can however be misleading. An example of this arises even in relation to the derivation of the name 'Tentsmuir.' It has popularly been claimed by some, that the name Tentsmuir comes from the adoption of the Scottish term *Densman* for a Dane. This has been used to argue that the name Tentsmuir can be interpreted as *the Muir of the Danes*.

A recorded account of this popular derivation is found in Boswell's *Biography of Samuel Johnson*. When Dr. Johnson and his biographer James Boswell came to St Andrews in August 1773 on their grand tour of north-east and north-west Scotland and the Inner Hebrides, they visited the ancient Norman Church at Leuchars (Figure 2.1). Here they met the minister whom they described as 'a very civil old man' who informed them that there was 'a colony of Danes in his parish and they had landed at a remote period of time and still remained a distinct group of people.' Dr. Johnson then asked if they had brought women with them. As no answer to this questions seems to have been forthcoming, Boswell then concluded 'we were not satisfied as to this colony' (Boswell, 1791). Even today it is sometimes asserted (see current SNH notice boards at Tentsmuir) that a Danish ship was wrecked on the coast and that some of the ship-wrecked mariners remained in the area giving their name in a proprietary fashion to this piece of coastal moorland.

The minister of Leuchars, the Reverend Robert Kettle writing in the Old Statistical Account in 1796, not long after the visit of Johnson and Boswell, states that '*Tentsmuiris* is a very large, flat part of a district on the east about which many wonderful stories have been told, concerning the original inhabitants and the peculiarity of their manners.' He states that 'after most laborious enquiry I find no reason to conclude, according to general report, that this part of the parish was peopled by the crew of a Danish fleet wrecked on the coast.'

In addition, the Reverend Kettle also endorses the word '*tent*' as referring to the tents lived in by the early shepherds, which could be seen by the inland inhabitants of the Parish meaning a portable shelter. The old French *tente* Latin *tenta* is attested from the 14th century, and *Muir* signifies rough grazing land. These portable shelters would therefore have been associated with transhumance activities on this large area of sandy pasture. The Reverend Kettle's remark about the visibility of the *tents* also indicates the open nature of the land and the absence of trees at this time.

It must therefore be concluded that to derive the name *Tentsmuir* from the arrival of Danish residents is without any foundation.

The Reverent Kettle also refers to Tentsmuir as S*heuchy Dyke* and describes it as a large tract of level, swampy, sandy ground, lying between the rivers Eden and Tay, which is intersected with ditches, which he calls *Sheuch*s, and *Dykes*, and where the *inhabitants build their houses of earthen sods or turf as no stones or materials for brick being found throughout the whole tract*. He also notes the presence of *lochs or lakes* of which he *says the one called the Canal Loch, seems to have been the longest and deepest* (see page 34).

The Normans at Leuchars

Tentsmuir and its surroundings were taken over by the Normans on their arrival in Scotland in the early 12th century. Norman personal names appear in Scotland during the last years of the reign of David I (1124-1153). They were few in number. One of the first to be recorded was 'Ness son of William' as co-signatory of a document which granted the Church of Leuchars to the St Andrews Priory in 1172. His baptismal name was Scottish not Norman, and it is therefore probable that he was the son of an early Norman immigrant and a Scottish woman and that he was born in Scotland before David came to the throne (Ritchie, 1954). Ness inherited lands in Fife, Perth and Lothian probably from his mother and as Lord of Leuchars appears to have lived there in considerable style with as many Norman Household officers as his neighbour the Bishop of St Andrews (Ritchie, 1954).

His daughter was named Orabile, which was a popular name at that time and occurred in the *chansons de geste* (songs of heroic deeds), based on medieval epic poems which were recited and sung by troubadours and reached their pinnacle of popularity between 1150 and 1250 (Hasenohr and Zink, 1992). The name Orabile (or *Orabilis*) sometimes given as Orabel is probably derived from the Latin *orabilis* –*orare* to pray i.e. to a saint who could be invoked. An alternative derivation may be connected with the fact that the name first appears in the Guillaume d'Orange cycle of *chansons de geste* that includes, as was often the case, names of exotic figures. Here Orable was a Saracen princess who married Guillaume and converts to Christianity. This would imply that Orable was a confabulated name, made up for the story, with *'Orab'* possibly derived from the French word 'Arabe' meaning Saracen (Hammond, 2013). The name in the form Arabella is still in use, but rarely found today.

Orable married three times. By 1170 she had married Robert de Quincy I whose father (Saher de Quincy I 1090-1158) took his name from Cuinchy near Béthune in the Pas de Calais. Robert de Quincy I, who had probably also been born near Béthune, settled around 1165 in Scotland pursuing a knightly career as a close companion of his cousins King Malcolm IV and King William I (Simpson, 1985).

At that time Scotland was considered as a land of opportunity by Norman incomers. When Orable confirmed the grant of the church at Leuchars to the St Andrews Priory *c.* 1185 she is recorded as *Comitissa de Mar,* wife of the Earl of that name.

An impressive and lasting visible testimony to the cultural influence of the Normans in Fife is found in the the Parish Church of Leuchars. This building with its romanesque apse and colonnade is first mentioned in a document granting it to the Priory of St Andrews, by Ness the son of William, the Lord of Leuchars with the approval of his daughter and heiress Orabile Countess of Mar *c.* 1185 (Taylor with Márkus, 2010). The Church in question was that of the mid 12th century Parish Church of Leuchars, one of the finest and rarest examples of Romanesque architecture in Scotland (Figures 2.1-2)

The current conception that the dedication of this medieval church of Leuchars to St Athernase is in fact an error (Taylor with Márkus, 2010). It arises from a mis-reading of a list of church dedications in the

Figure 2.1 The ancient Norman parish Church of Leuchars with its 12th century Romanesque apse and chancel now popularly called St Athernase due to a clerical error (see text). The bell tower was added in 1745 and the nave in the 1800s.

Tentsmuir in history 15

Figure 2.2 Leuchars Church (St Athernase - see text) as it was on 18 July 1832 with the original nave and mullion windows (adapted from Guide and History; St Sylvia Cassells 1995).

Register of St Andrews Priory in a medieval manuscript, now in the National Archives of Scotland which lists the churches dedicated, or re-dedicated, by Bishop David de Bernham of St Andrews in the 1240s.

The eighth church in this list is *'ecclesia sancti Johannis evangeliste et sancti Athernisci confessoris de Losceresch'* (the church of St John the evangelist and St Athernase the confessor). However, the church of *Losceresch* is not the church of Leuchars, which in medieval sources is spelt *Lochris, Locres* etc., but the parish church of Lathrisk (now Kettle Parish in Fife), where the early spellings are *Losresc* (1170s), *Loseresch, Losseresc* (1227). Athernase is therefore the patron saint not of Leuchars but of Lathrisk. It is not certain what the original dedication was of the parish church at Leuchars. A chapel is mentioned along with the Church in the mid-13th century dedicated to a St Bonoc(us) and later variously to Nennet, Bernard and Bunyon but this was probably on a site near the present school. From an historical point of view, it might be more correct to refer to the Parish Church of Leuchars as St Bonoc or perhaps St Bunyon (Talyor with Márkus, 2010). The last Chaplain of St Bonoc held some extra posts in St Andrews, including a chaplaincy at St Salvator's College (Smart, 2015).

The building of the Leuchars Church probably began in the mid 12th century when Ness son of William inherited his mother's estates. The construction was probably still in progress when his daughter inherited her father's lands and married Robert de Quincy who through this marriage to Orabile, became lord of an extensive complex of Scottish estates which included lands in Fife, Strathearn and Lothian.

Robert de Quincy is credited with the building of Leuchars Castle located where the Motte at Leuchars is situated (Figure 2.3). This Motte had probably been constructed for some earlier fortification, most likely in the 12th century. Robert de Quincy may not have seen this work finished as he died in Palestine in 1197 AD. His son, Saher de Quincy IV, married the daughter of the Earl of Leicester. He led a busy and important public life, served both Richard I and John and was elevated Earl of Winchester about 1206-7. Earl Saher, like his father, also became a crusader and died in Egypt at Damietta in 1219 (Simpson, 1985).

After the death of Saher de Quincy IV, his son Roger acquired lands both in Scotland and England and on his mother's death inherited further estates in England as well as the title of Earl of Winchester. He married

Figure 2.3 The mound now known as Castle Knowe is all that remains of the former medieval Motte and Bailey Castle built at Leuchars in the late 12th century. The Motte at present has a height of 25 feet. Earlier measurements record a higher figure but this has ben reduced as a result of agricultural activity. By the early 13th century as a result of de Quincy marriages (see text), Sir Roger de Quincy could ride out on the road shown above in the foreground and travel as far as Dover, and apart from a 100 mile stretch in the north of England, would never have been more than 30-40 miles from land in which he had an interest (Grant Simpson 1985).

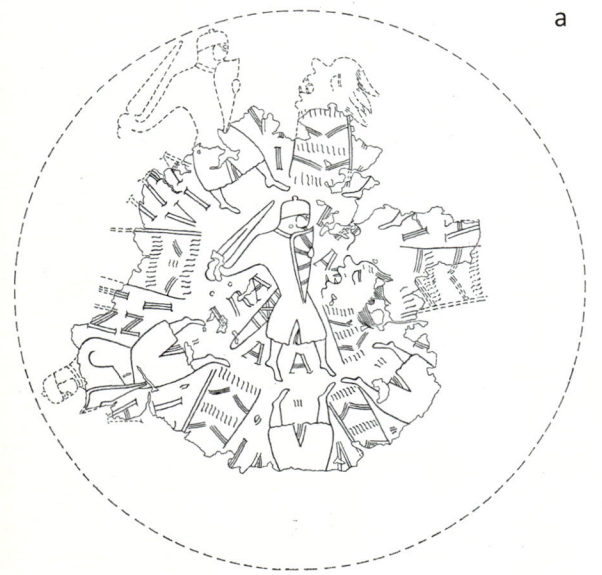

Figure 2.4. (a); drawing of reconstruction of part of an Engraved Norman Bronze bowl found in 1923 in ploughed land to the south of the Castle Mound (the Motte - Figure 2.2) upon which Leuchars Castle once stood. Only the central portion of the plate was found as shown here (b) and exhibited in the National Museum of Scotland. It appears to be of a type that was common in North Germany and described as *Ritter Schalen* and is similar to the more complete examples in London Plate and Kölner Schale (see PSA 1931). A Knight in armour, helmet and vizier fights a monster. The letters IRA and AA appear twice. (Source of drawn image – Fife Inventory drawing).

three times. The first of his three wives was the rich heiress *Helen of Galloway* and through this marriage he also obtained the honorific title of *Constable of Scotland*. His second wife *Maud de Bohun* was the daughter of Humphrey de Bohun, 2nd Earl of Hereford. Maud died only two years later, and Roger in the same year married his third wife, Eleanor de Ferrers, daughter of William de Ferrers, 5th Earl of Derby.

Notwithstanding this great wealth, and despite extensive travel as ambassador for Henry III, his political career has been described as undistinguished (Simpson, 1985). He had no male heir to succeed him. His first wife Helen of Galloway bore him three daughters but there was no issue from his later marriages. When he died on the 25th of April 1264, as the last male of his house, he would have been close to seventy, and lacking a son would have been a very disappointed man (Simpson, 1985).

An indication of the wealth and sophistication of this northern Norman household is indicated by the discovery of a fragment of a very damaged copper-alloy hammered and engraved bowl found in 1923 by Dr. J.B. Mears rolled up and buried in a ploughed field to the south of the Motte at Leuchars (Figures 2.3-4).

The engraving has been interpreted through comparisons to other similar bowls as showing a lively figure of a knight emerging from a tree stump and wielding a sword against a monster's head. This is a depiction of the battle between good and evil with two of the seven deadly sins (pride and anger) spelled out in capital letters among the trees (Crawford, B. 2017). It was probably made in the Rhineland in the mid-12th century (Glenn, 2003).

The first explicit reference to Leuchars Castle (Figure 2.5) occurs between 1306 and 1329 when King Robert I divided the lands of Leuchars into three distinct baronies. These were subsequently known by the family names of the feudal superiors. Thus, the part of what had been the Lordship of Leuchars and contained the Castle, became known as Leuchars-Ramsay. This title with the relevant charter first appears in February 1380. The charter included specifically 'our castle at Leuchars' (*castrum nostrum et omnes et singulas terras nostras cum pertinentiis iacentes ei Baronia de Locrys*).

Just how much of the castle remained at that time is not altogether clear. The castle had been stormed, taken, and demolished by the Earl of Pembroke in 1336. It was then ordered to be rebuilt by Edward III. How much

Figure 2.5 Leuchars Castle from a drawing in 1785 by William Crawford before its demolition in the late 19th century (Reproduced with permission from Handbook of Leuchars Parish Church by William Borthwick - St Andrews Citizen Shop 1954).

was rebuilt is somewhat doubtful as it was razed to the ground again in 1337 by the Scots under the command of Andrew of Moray, Guardian of Scotland. Despite this period of turbulence, the castle is mentioned again in 1376 in a grant to William Ramsay.

Restoration of Leuchars Castle was undertaken by various owners and was still occupied in 1565 when it was described as having been founded 'on a broad motte including 3 acres of ground with an apartment of 6 rooms with marble chimney pieces and wainscoted with oak and belonging to the Earl of Southesk' (Gifford, 1988). It may even have hosted royal visitors as Tentsmuir had been designated as a Royal Hunting Forest by James III in 1468. The castle was finally demolished in the late 19th century by the Earl of Crawford at a time of extensive agricultural improvements.

Late medieval Tentsmuir

After the death of Roger de Quincy in 1264 his estates were divided between his three daughters. The first daughter, Elizabeth (also known as Isabel), married Alexander Comyn, 2nd Earl of Buchan. This period in the 13th century may have seen the first use of the name 'Earlshall' for the family home, as after the death of her husband Elizabeth, then Countess of Buchan, changed the name of her residence to *Countess Hall* (Talyor with Márkus, 2010).This would suggest that during her husband's lifetime and afterwards, the family home was known as Earlshall.

There is also a reference to a building called Earlshall in 1516 before the construction on the site of the present Earlshall which was begun in 1546 by Sir William Bruce. A charter of James V dated at Earlshall in 1540 confirmed the sale of various lands to William Bruce of Earlshall. This suggests that there was a property that had survived with the name of Earlshall possibly via Elizabeth Countess of Buchan. According to RCAHMS the erection of the present Earlshall was begun in 1546 and the south range added in the 17th century (Figure 2.6) The building became ruinous about 1890 (Taylor with Márkus, 2010).

In 1891 this outstanding example of a Scottish Tower House was purchased and rescued by the linen manufacturer R.W.R MacKenzie of Stormontfield who then charged Sir Robert Lorimer with the restoration. Lorimer's restoration (Figure 2.7) included the addition of a gatehouse and some unusual topiary gardens which include chessmen. Lorimer also restored the 16th-century painted ceiling in the Long Gallery which depicts subjects from heraldry, history, and zoology (see Fife Inventory).

Figure 2.6 Earlshall before restoration from an engraving by William Ballantyne, 1872. (Reproduced with permission from Adamson, P. *Early photographs of St Andrews*

Tentsmuir - a royal hunting forest

Pitlethie lies close to the site of Leuchars Castle and is first mentioned as church land in the 12th century. St Andrews Priory built a high status residence here in the later Middle-Ages. Both James V and VI used Tentsmuir as a royal hunting forest. In medieval terminology a *forest* was a region belonging to a king or nobleman with its own laws. James V issued at least 16 charters from Pitlethie between August 1537 and May 1542. Such was the frequent royal use of Pitlethie that it was sometimes referred to at that time as *the Palace* (Taylor and Márkus, 2010e).

Salmon fishing rights

Royal authority is also evident in relation to the rights over the various fishings which until relatively recent times have been one of the most significant financial resources of Tentsmuir. The different sections of the shore were therefore clearly named and duly recorded. In 1539 James V granted to his treasurer James Kircaldy of Grange and his heirs the fishings (*piscaria*) on the Tay lying next to the lands of Fetters and Shanwell (Gaelic – Old Farm *Sean baile*) begining at the Lundin Burn on the north side of Shanwell and going down to the Pool at the Powie Mouth (Taylor and Márkus, 2010e).

The Powie Burn (Figure 2.8) is cited as a marker for this fishing right that stetched from the north side of Shanwell going down by fishings at Green Scalp and Lucky Scalp to the Pool at Powie Mouth (Taylor and Márkus, 2010f). The term Scalp probably refers to mussel beds. *Pow* is the term for a slow-moving stream.

Figure 2.7 View from the west of the restored medieval Castle of Earlshall (Photo author).

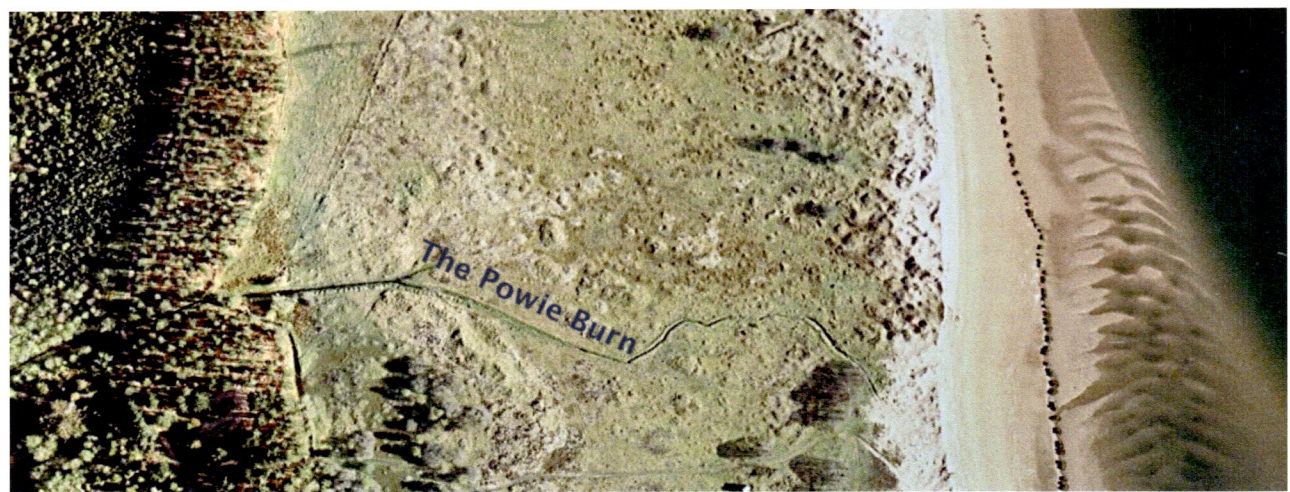

Figure 2.8 The Powie Burn as it disperses across the sand dunes with only a very minor section reaching the shore: as seen from a satellite image

The reference to the Powie Mouth in the grant of fishing rights in 1539 indicates that the stream reached through the dune system to the sea. By the 1900s this was no longer the case. Today the Powie Burn is little more than a minor ditch usually found disappearing into the sands at a considerable distance from the sea and with only a minor section visible on the landward side of the last line of dunes.

This blockage of open, surface drainage to the sea by the growth of the sand dunes has maintained the tendency for extensive areas of wetland and marsh to dominate much of Tentsmuir away from the front line of dunes.

The Salmon fishing rights as well as being clearly laid out in documents dating back to the 16th century were also marked out on the ground.

A boundary marker erected at the end of the 18th century still stands today as an inscribed stone According to the inscription on its north side (now no longer readable, but see photographs taken in the early 1960s), which states that on the south side of the stone that '*This stone was erected in the year 1794*' (Figure 2.9 a-c).

To be able to project a straight line from the top of Norman's Law using this marker down to the low water mark would have required this stone when it was erected in 1794 to have been on or near the top of the front line of dunes which would have been reasonably close to the sea. The stone is therefore an indicator of the approximate position of the front line of dunes and where the coastline lay in 1794. The stone is now just over 1 km from the sea. In 1941 when the anti-tank blocks were set up it was only 650 m from the sea (see Chapter 3).

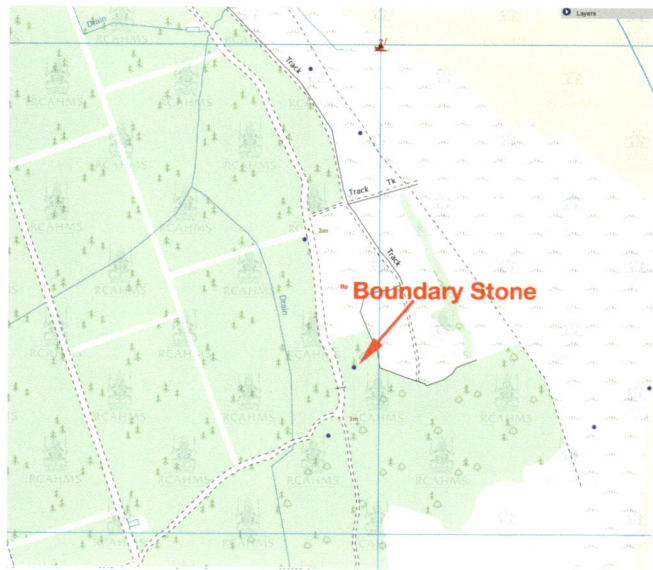

Figure 2.9 The Salmon rights boundary stone (a) Grid Reference NO499272. The inscription on the north face of the stone reads: The march between the Shanwell and Old Muirs Salmon Fishing is a straight line from the top of Norman's Law to the low water mark. This stone stands in the said straight line.' The date of erection (b) is given as 1794. The location of the stone (c) now on the edge of the forest as shown on OS map 2014.

Salmon fishing from fixed nets has long been carried out at Tentsmuir with bothy accommodation for the manning and servicing of the nets round the clock depending on the tides. The brick foundations of the fishermans' bothies were still visible in the 1960s. The advent of the railways increased the demand for fresh salmon and the Icehouse which was originally built in 1855 was extended later in 1886 as recorded on the lintel over the door to the lower section of the building (Figure 2.10).

Agricultural improvement

As already discussed, Tentsmuir has long attracted human settlement. One favourable factor would have been that it is that it is one of the sunniest places in Scotland (Table 2.1) It also enjoys on average an evenly monthly spread of rainfall (Figure 2.11).

Tentsmuir avoids excessive precipitation and cloud cover from its situation on the lea side of mountains. Consequently, a warm *Föhn* situation can develop. This happens most frequently when winds come from a westerly or north-westerly direction becoming warmer and drier as a result of their water vapour being left behind on the high ground to the west.

Despite Tentsmuir being largely composed of sand, enjoying the best sunshine regime in Scotland, and only moderate rainfall, it has always had areas prone to flooding. This was already apparent at the time of the Mesolithic activity in the early Holocene. In the medieval period considerable stretches of land around Leuchars were dominated by lochs and bogs. An example of this was *Rash Myre* which lay between Leuchars and Morton and is depicted in the Bleau Atlas of 1654 (Figure 2.12).

Figure 2.10 The Tentsmuir Icehouse – built in 1855 with an extension added in 1886. Snow-ice would have been collected locally but ice was also shipped in from the Baltic and the lower extension added as the export of salmon was facilitated by the development of rail connections.

Tentsmuir in history

Dundee	1564
Leuchars	1563
Tiree, Inner Hebrides	1477
Aberdeen	1433
Edinburgh	1421
Forres	1307
Glasgow	1265
Stornoway, Outer Hebrides	1224
Oban	1219
Kirkwall, Orkney Islands	1172
Lerwick, Shetland Islands	1110

Table 2.1 Average hours of sunshine at various locations in Scotland, 2013-2019.

Figure 2.11 Average monthly rainfall at Leuchars, 2013-2019.

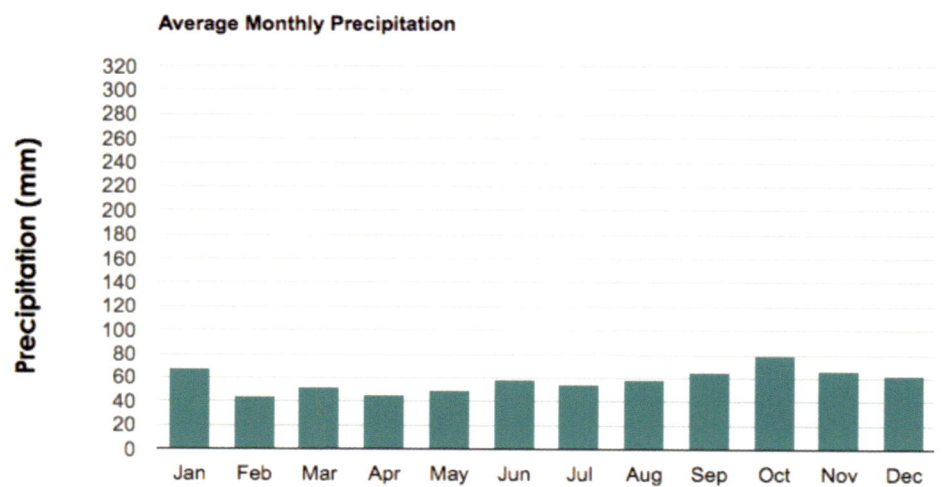

In the southern part of Tentsmuir there were also several long narrow lochs running almost due north to south in the hollows of the dunes. These long narrow lochs would have developed from former dune slacks (*winter lochs*) that form between successive lines of dunes (see Chapter 5).

As already pointed out, the sandy nature of the soil and the relatively low rainfall in Fife has not prevented Tentsmuir from having problems with drainage and flooding. This has been a feature of the region for over 10,000 years. Even today where drains have been dug the fluctuation of the water table between summer-lows and winter-highs poses problems for agriculture and forestry. In winter, the water table can be too high for pine, and in summer too low for spruce.

The draining of Tentsmuir

Already in the 16th century it was recognised that flooding was the greatest impediment to agriculture at Tentsmuir. The sandy nature of the soil facilitated the concerted efforts of the early agricultural improvers. Possibly the earliest attempt to improve drainage was the creation of the Canal Loch - a name that was recorded in an early charter from 1540 indicating an organized attempt to improve drainage at Tentsmuir (Taylor and Márkus, 2010i). The location of this loch, erroneously named as Candle Loch, is shown in the south-east corner of Tentsmuir in the Bleau Atlas of 1654 (Figure 2.12).

These permanent lochs were further drained in the eighteenth cenutury by a series of canal-like waterways which are notable for their length and straightness (Figure 2.13). This drainage activity was described by the minister the rev. Kettle writing in the mid 1790s as four long, broad and beautiful and almost parallel canals. In the year he is writing he records that '*the tenants are cutting drains (casts) and letting the water out of these canals* 'to render the pasture more beneficial to their cattle' (Kettle, 1796)

Tentsmuir: Ten Thousand Years of Environmental History

Figure 2.12 Tentsmuir and its wetlands as depicted in the Bleau Atlas of 1654

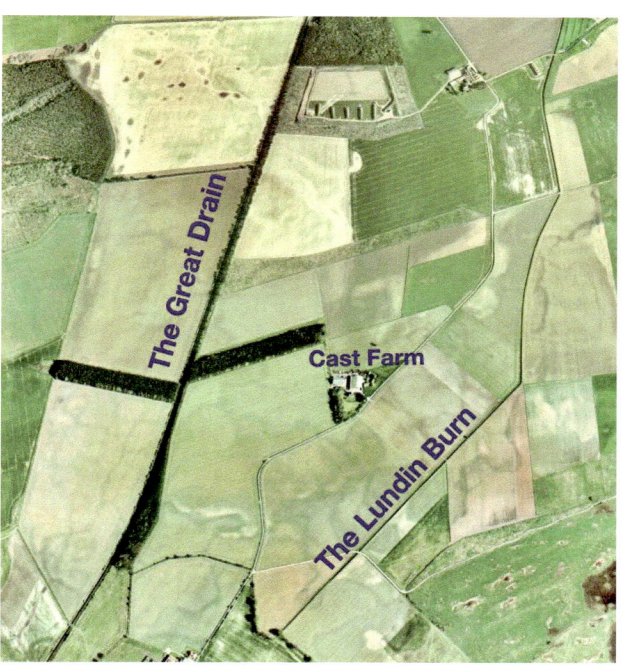

Figure 2.13 The Great Drain constructed in the latter part of the 18th century.

The drainage activity that had already been pursued at this southern edge of Tentsmuir created the farm that still carries the unique name of *'Cast Farm' - the farm between the two ditches.*' It is first mentioned in 1785 when the tenant was Archibald Johnston. The western drainage ditch would appear to be the Great Drain (Figure 2.13) mentioned as having been dug by Sir David Carnegie in 1785 before he sold the lands to Robert Lindsay (Taylor and Márkus, 2010)

The land around the abandoned medieval castle was also a swamp which was only improved with the digging of the Great Drain which took water to the river Eden via Leuchars and added many acres of fertile land which were reported as producing all kinds of grain, clover, turnip and cabbage where before there has only been coarse grass and rushes (Kettle, 1796).

Malaria at Tentsmuir

The drainage of Tentsmuir not only enhanced agricultural productivity but also improved the health of the local people whom are recorded as suffering from intermittent fevers from spring to autumn (Taylor and Márkus, 2010) (pp 483-4). The unhealthiness of coastal and wetland environments had long been associated with 'marsh fever' or 'ague'. Before the infectious nature of the disease had been discovered it was attributed to the environmental condition the *mal-air*. Malaria was widespread in Britain and there had been unusually high levels of mortality from the sixteenth to the nineteenth century particularly in marshy areas.

Studies of the geography and history of the 'marsh fever' have shown that in fact, malaria in Britain was transmitted by anopheline mosquitoes. However, the species of malaria organisms endemic in Britian in the past were *Plasmodium vivax* and *P. malariae* both of which caused the so-called 'benign' forms of malaria and were not the more deadly tropical strains of *P. falciparum* (Dobson, 1994).

It is thought that the benign forms of malaria were introduced to Britain during the Roman occupation (AD first to fifth centuries). However, the lack of written mortality records prior to the post-medieval period has made it difficult to evaluate either the extent of the presence or impact of the disease. Incidence of mosquito-borne disease in Britain has been strongly influenced by land use, and lifestyles affecting human-mosquito contact. Since this was discovered and remedial measures implemented there has not been a single human case of a mosquito-borne virus detected in Britain for almost a century and a half, (Ramsdale and Gunn, 2005). Leuchars is believed to have been the last place in Scotland to become free of Malaria.

Tentsmuir's farms

With the draining of Tentsmuir agriculture flourished. A rapid increase in arable farming replaced grazing as the main agricultural activity. Possibly the most notable farm to prosper was Craigie (see Figures 2.14-15). This farm has long been in existence. There is a record of land being added to this farm by William de Ferrars Laird of Leuchars granting additional land to Robert de Harcars *c.* 1300 (Taylor and Márkus, 2010).

By 1785 Craigie was already a large farm and had become the second largest farm in Fife second only to Kincaple which lay on the southern side of the river Eden (Figure 2.14), and carried 12 pair of horse (Taylor and Márkus, 2010f). The number at Kincaple rose subsequently to 15 pairs (MacLeod, 1996). A well known Bothy Ballad entitled *Tatttie Jock* composed in the early nineteenth-century about a farmer John Mathie of Craigie where 5 of his ploughmen whom he reported to the police for stealing his potatoes from his shed were banished to Australia for 14 years. One however avoided transportation by enlisting. This event gives an insight to social conditions under which farm workers served at that time. John Mathie himself was distressed about their banishment, as even his offer of payment of 1,000 guineas to the authorities did not save their fate.

One note of comfort that is recorded, is that one of the men on his return from banishment, went back to the spot where he had hidden some money etc., under a stone which none but he could lift and happily found that the money was still there (MacLeod, 1996)!

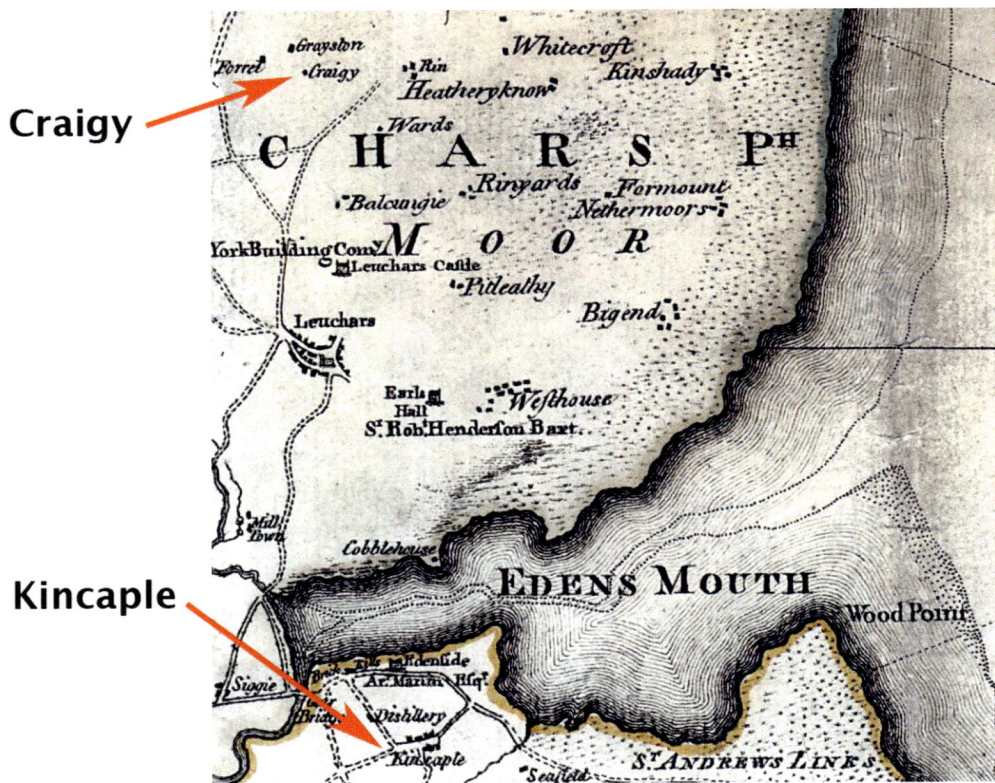

Figure 2.14 Ainslie map of Tentsmuir 1775 showing the principal farms in the southern and more fertile portion of the Tentsmuir Peninsula. (Reproduced with permission from the National Library of Scotland)

Tattie Jock

From the singing of Fife traditional singer Archie Webster of Strathkinnes. A Bothy Ballad with a transportation theme published in broadsheet form.

Ye'll a' hae heard o' Tattie Jock
Likewise o' Mutton Peggie.*
They had a fairmie ower in Fife,
An' the name o' it was Craigie.

Chorus
Singin' ah riddle aye
Roo dum di do.
Ah riddle aye
Roo dum day.

There was ten pair upon that place,
Likewise ten able men,
It's five they gaed for tae kinnle the fire
An' the ither five oot tae scran.

Three month we served wi' Tattie Jock,
An' weel we did agree,
Till we found oot that the tattie shed
Could be opened wi'the bothie key.

We a' went intae the tattie shed
Oor bags were hardly fu'
When Tattie Jock in ahint the door
Cried 'Aye, ma lads, stand still!'

* *Mutton Peggie was John Mathie's housekeeper*

Oh, the first he got was Willie Marr,
The next was Sandy Doo,
There was Jimmy Grey an' Wull Moncur
An' Jimmy Pethrie flew.

Next day some o' us were drivin' dung
An' some were at the mill,
The foreman he was at the ploo
Upon Pitlootie Hill.
They sent for ten big polismen
But nine there only came
It dinged them for tae lift's that nicht
Us bein' ten able men.

The hin'maist lad was the wisest een,
The best lad o' us a'
He jined a man o' war at Leith,
So's he didnae need tae stand the Law.

When we were gettin' oor sentences
We a' stood roond an' roond
But when we heard o' the fourteen years
Oor tears cam' rollin' doon.

When Tattie Jock heard tell o' this
He cried and grat fu' sore
A thousand guineas he would pay
If that would clear oor score.

A bag o' gold he did produce
Tae pey it there an' then
But the lawyer only told him money
Wouldna clear his men.

An' when they mairched us up through Perth
We heard the news boy say
'It's hard tae see sic able men
Rade aff tae Botany Bay.'

When we arrive in Botany Bay
Some letters we will send,
Tae tell oor freens the hardships we
Endure in a foreign land.

The name of the nearby farm of Rhynd (Old Gaelic *rind* -modern Gaelic *rinn*) denotes a point or a headland which reflects its position at the edge of the headland to the east of Craigie (Taylor and Márkus, 2010m). To this day the soil on these farms in Tentsmuir is subject to flooding due to deposits of clay (Figure. 2.15) probably due to its ancestral location near the former wetland of Rash Myre.

Arable land decreases towards the east of Tentsmuir and is replaced by less productive heathlands. This may explain therefore the origin of the name *Comerton* of the neighbouring farm to the east. The prefix *comer* meaning *difficult to work* (cf German *Kummer* = trouble, problems, grief, sorrow).

A feature of at least two of the farms nearer to the coast was the former presence of watch-towers. Comerton had such a tower as did also Kinshaldy. It is remembered that these farms also enjoyed salmon fishing rights and these towers were used as lookouts against possible poachers (information from a descendent of the 19th century tenant farmer at Kinshaldy).

Both Duck and Grouse shooting took place at Tentsmuir from the latter years of the 19th century

Figure 2.15 The farm of Craigie in southern Tentsmuir. By 1785 this was the second largest farm in Fife (see text) and the scene of the deportation of 5 ploughmen to Australia in the early 19th century for stealing potatoes. This image taken in February 2015 also shows a drainage problem due to clay deposits probably due to the proximity of this land to the ancient wetland of Rash Myre (see Figure 2.12).

up to 1919. For a short period the northern part of Tentsmuir was managed as a grouse moor under the tenantry of Mr. William Berry (4th of Tayfield) who also wrote an account of the introduction of grouse (Berry, 1894).

During the 1914-18 war most of the old pine woods of Tentsmuir were felled leaving just a few isolated old pines. The shootings changed hands in 1919 when the estate was purchased by the Town Council of Dundee for possible industrial use including the pre-fabrication of parts for ships (see p 95). This however did not take place and the greater part of the moor was bought by the Government for afforestation.

The extreme southern end of Tentsmuir, with its low lying level land together with its outstanding cloud-free weather, has also served for the defence of the realm as for a century it was an aerodrome for the Royal Air Force from 1915 up until April 2015.

Up until the 1930s the Kinshaldy area of Tentsmuir was commonly used for camping, both tents and for some, even horse-drawn caravans. It is therefore no surprise that for some families these were prolonged summer visits for mother and family living often in caravans which father would visit at weekends.

German spies at Tentsmuir

There are also records of German espionage at Tentsmuir and elswhere in Fife in the years runnng up the beginning of the Second World War. In May 1932 two Germans science students arrived, that for some reason had to be accommodated in Deans Court, a fine 16th century house in St. Andrews, which at

Figure 2.16 Prime Minister Winston Churchill and General Władysław Sikorski, Commander in Chief of the Polish Armed Forces & Premier of the Polish Government in Exile, inspecting troops of the Polish 1st Rifle Brigade engaged on the construction of coastal defences at Tentsmuir on the 23rd September 1940. (The St Andrews Citizen)

that time had just begun to be used as a University residence. They claimed to be making a study of the algae of the harbours of Fife. They then disappeared as mysteriously as they had arrived. At that time Nigel Irvine, the son of the University Principal (Sir James Irvine) had built himself a hide for watching birds at Tentsmuir. One day he found there a map of the whole area that surrounds the territory of RAF Leuchars. On it were marked in German the flight paths and other details at this aerodrome of this then little known corner in the East Neuk of Fife (Melvin, 2011).

It was therefore clear to the military authorities that Tentsmuir was a likely landing place for a German invasion of Britain. Soon after the beginning of the 2nd World War an extensive line of concrete blocks was constructed along the shore for the entire distance between the estuaries of the Eden and the Tay. For this task the Polish soldiers who had escaped from the onslaught of the German army were given a camp site with kitchens in the Tentsmuir Forest the remains of which are still visible.

A railway line was constructed and equipped with bogies to transport the blocks once they were made. They were also adapted to provide moving targets for training RAF Spitfire pilots. On the 23rd September 1940 Prime Minister Winston Churchill and General Sikorski visited Tentsmuir and inspected the Polish troops engaged on this urgent defence work (Figures 2.16-17).

The anti-aircaft gunners at Tentsmuir were also given specific instructions, that when firing at German aircraft, they should aim to miss the spotter-planes so that they could return to base and report how well Tentsmuir was defended!

Figure 2.17 Trolley as used for holding targets that moved along a railway to provide shooting practice for RAF Spitfire pilots (Photo T. Cunnningham).

Chapter Three

Sand and water

Sand and water interactions

Sand varies in its physical relation to water. On one hand, sand in the soil aids drainage, while on the other hand compacted sand can cause flooding by impeding water movement. This is evident in the use made of sand bags for domestic flood protection.

Across the Tentsmuir landscape these divergent properties of sand are revealed in alternating patterns of dry dunes and intervening wetlands, commonly called *dune-slacks* or *winter lochs*. The soil in these wet areas becomes compacted causing the water table to rise. This is particularly prevalent in the colder periods of the year when evaporation from the soil surface and transpiration from the vegetation is reduced. The further the distance from the foreshore, the greater is the increase in lateral resistance for drainage to seaward in the compacted sand. Consequently, it is the landward slacks that are most prone to flooding.

This duality in the nature of sand in relation to water, stems from the unique properties of sand grains. Sand is neither solid nor liquid. It is granular, and therefore capable of behaving as a free running material, easily blown by the wind and percolated by water. Due to properties of cohesion between sand grains, sand can accumulate to form dunes as well as spreading out into plains. When permeated by water sand loses its cohesive properties between grains and can become compacted thus impeding drainage (Figure 3.1).

Free moving sand can demonstrate an impressive capacity for self-accumulation, especially when it builds itself into dunes. Depending on conditions, dunes can be just a few metres high as at Tentsmuir.

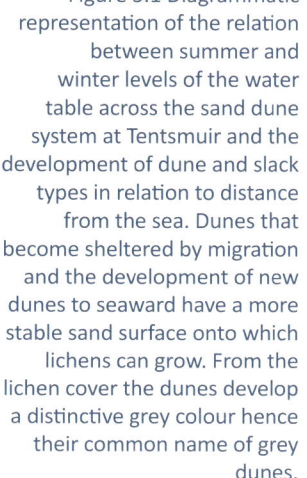

Figure 3.1 Diagrammatic representation of the relation between summer and winter levels of the water table across the sand dune system at Tentsmuir and the development of dune and slack types in relation to distance from the sea. Dunes that become sheltered by migration and the development of new dunes to seaward have a more stable sand surface onto which lichens can grow. From the lichen cover the dunes develop a distinctive grey colour hence their common name of grey dunes.

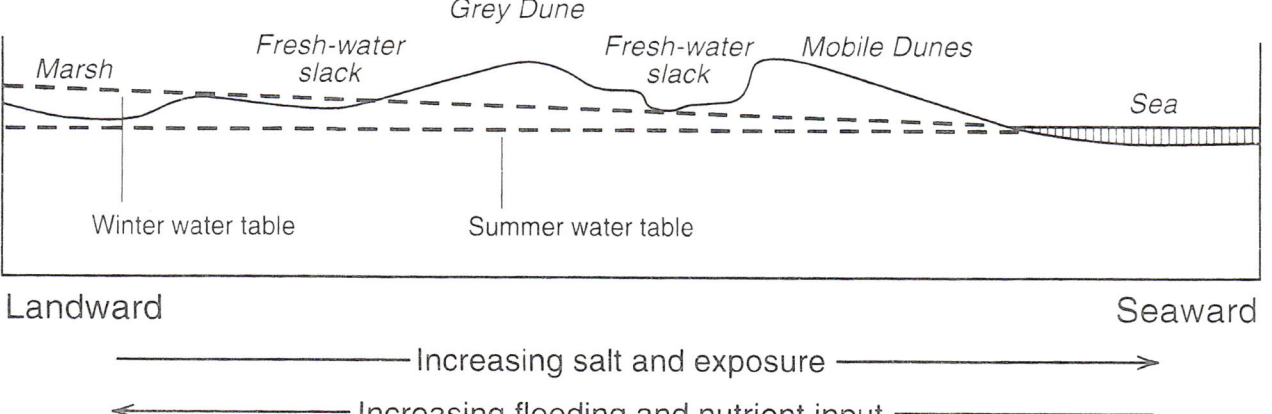

Much larger dunes however can be seen in the Baltic. The most outstanding example of which is the 98 km long Curonian Spit which lies with its northern part in Lithuania and its southern portion in Russia (Figure 3.2-3). This unique sand-spit has the highest moving (drifting) sand dunes in Europe with an average height of 35 m., and some portions even attaining a height of 60 m. (Bagdanaviciute et al., 2015).

Whether it be Tentsmuir, or the Baltic, sand dunes are prone to damage when disturbed, particularly when the sand is not anchored by vegetation. In the high sand dunes of the Curonian Spit it is has been estimated that one person walking over a dune can displace as much as 10 tons of sand (Lithuanian Conservation Notes Klaipéda, 2014). Similar damage on a smaller scale can also take place at Tentsmuir. Although the tonnage of sand displaced is less, the relative damage to the dunes can be just as severe.

Studies on the Curonian Spit have been used to determine the factors that affect the vulnerability of sand dunes to disturbance. It would appear that it is not just exposure to wind and waves that affect sand dune fragility. Just as important, are factors such as the rate of shoreline change, beach width/height ratio, underwater slope, and offshore sand bars, (Bagdanaviciute et al., 2015).

When free of vegetation, dune systems are prone to migrate in relation to the direction of the prevailing wind. In this manner, sand can move with great rapidity when triggered by a suitable set of circumstances. Even when dunes have been stationary for many years they can suddenly move *en masse*, as has been demonstrated time and again when whole dune systems were either removed in storms in a matter of days, or else migrated inland to cover older established soil surfaces. Such an event took place during a windstorm in 1694 in Scotland at the Culbin Sands. After this event, the Culbin Sands remained a region of mobile dunes for two centuries, causing the region to be described as *Scotland's desert* (Ross, 1993).

Similar events have also taken place in the past at Tentsmuir and remnants of these movements can still be seen in in the open moorland in the south-east part of

Figure 3.2 View of part of the Curonian Spit (Lithuania) a 98 km long, thin, curved sand-dune spit that separates the Curonian Lagoon the from the Baltic Sea. In places the spit is over 60 m high (Photo author).

Figure 3.3 The 98 km long Curonian Spit stretching from Lithuania in the north to Russia in the south

Figure 3.4 Ancient dunes towards the southern end of Tentsmuir showing an early sand deposit over which a former soil horizon developed and onto which a later dune then migrated.

the Peninsula. Here, ancient sand dunes have migrated over a fixed organic upper soil horizon which in turn had formed over a more ancient dune (Figure 3.4)

Tentsmuir provides striking examples of several dune types which vary with age and the nature of the vegetation cover. These are discussed in relation to their biodiversity and vegetation in Chapter 4.

Afforestation of Tentsmuir

In the early 1920s the greater part of northern Tentsmuir was bought by the Government for afforestation (see Chapter 1). This newly purchased land was then planted with trees up to the front line of dunes. The location of the front line of dunes at that time can therefore be seen from the seaward extent of this c.1924 tree planting (Figure 3.5).

The onset of the Second World War necessitated the emplacement of coastal defences on many coastal sites in Britain. This included Tentsmuir, where a line of anti-tank blocks were laid to seaward of the front line of dunes in 1940. This defence-line therefore serves today as an indicator of the position of the shoreline at that date (Figure 3.6) The photograph below taken in 1957 shows the seaward limit of the 1924 tree planting, as well as the high water mark in 1940, as this was the line for placing the coastal defence anti-tank blocks in order to hinder any German coastal invasion. It is therefore possible to trace the development of Tentsmuir Point and measure the accretion that took place between 1924 and 1940 and also from 1940 to 1957 when this photograph was taken. The extent of the subsequent growth up to 1963, with the further development of the Great Slack in the ensuing 6 years is shown in Figure 3.7.

In the early years of the 1939-45 war a concrete arrow and watch tower were put in place on the front line of dunes to direct and observe aerial bombing practice on off-shore targets (Figure 3.7). This image also shows the position of a flood-line of alder trees which had become established on the shore-side of the first dune slack. The alder seeds (*Alnus glutinosa*) tended to wash up at the edge of flooded areas where they germinated more readily than on ground which was actually flooded. Under such conditions they typically form what is described as a *Flood-Line Alder Association* (Figure 3.7).

The lighter coloured dune to landward is a typical grey dune. The characteristic feature of these dunes is the stability of the dune surface which was sufficiently stable to allow the colonisation of slow-growing lichens (see Chapter 5).

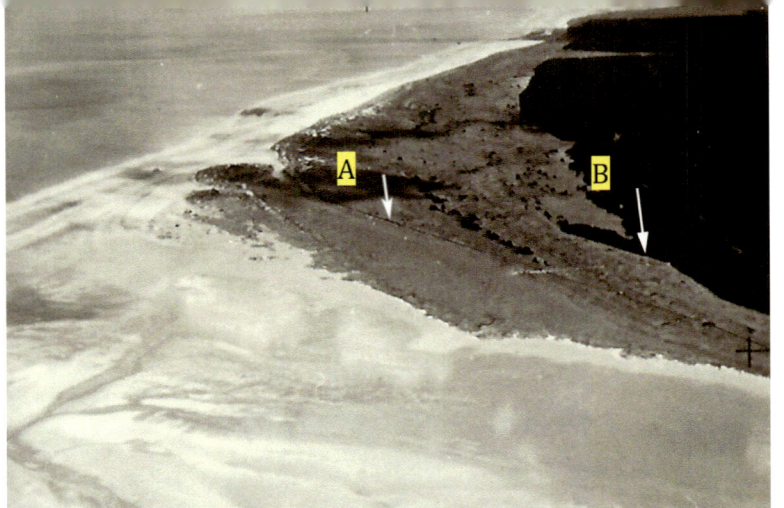

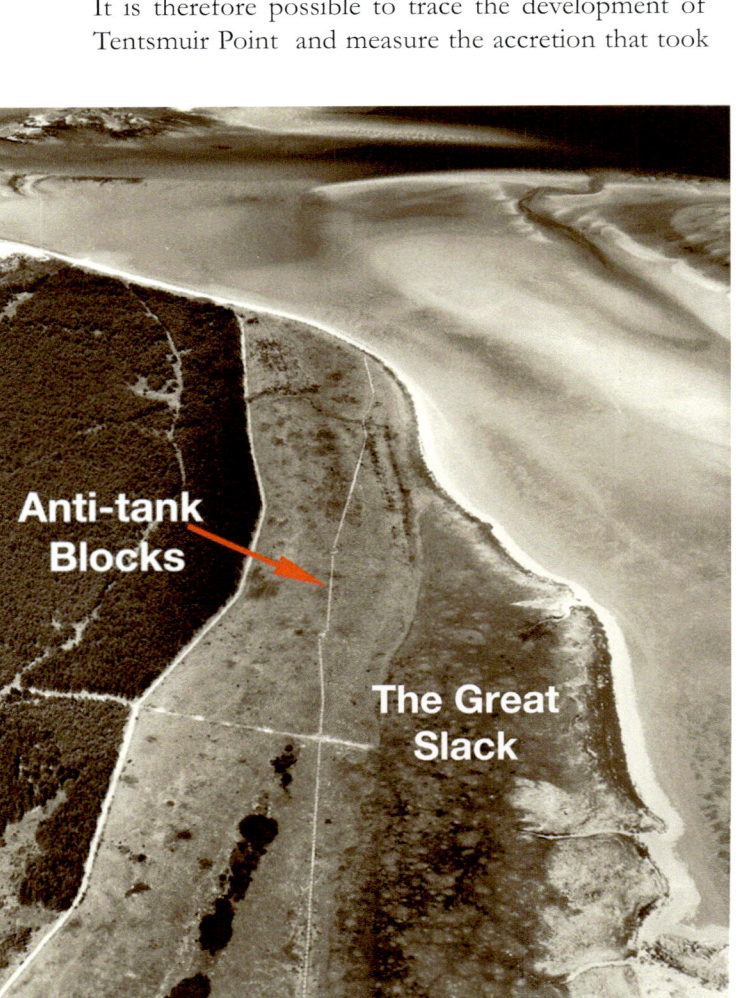

Figure 3.5 Aerial view looking south over the Tentsmuir Nature Reserve in 1957. Arrows denote (A) line of anti-tank blocks placed on high-tide level in 1940 and (B) the tree line planting limit in 1924 on the front line of dunes. Note minimal dune development at the northeast extremity of the Reserve (Photo courtesy of Dr. J.K. St Joseph – Cambridge University Committee for Aerial Photography).

Figure 3.6 View of the north end of Tentsmuir taken in 1957 showing the line of coastal defence anti-tank blocks that were put in place in 1940 on the high-tide line and the advance of the coastline that had taken place to seaward in the ensuing 23 years. (Photo courtesy of Dr. J.K. St Joseph – Cambridge University Committee for Aerial Photography).

Figure 3.7. Development of dunes and dune slack as photographed in 1957. Flooding of the slack deposits Alder seeds (*Alnus glutinosa*) along the upper limit of the flood line of the slack. To landward the slack is bordered by a grey dune. (Photo courtesy of Dr J.K. St Joseph – Cambridge University Committee for Aerial Photography).

Figure 3.8 Aerial view looking northeast over the Tentsmuir Nature Reserve in 1957 showing dune and slack series. (Photo courtesy of Dr J.K. St Joseph).

Dune slacks

The flood-plains found between the dunes are termed *slacks*. The term slack as used here, has a separate etymology from the use of the word to describe looseness. The term *slack* in relation to dune systems, means to become hydrated (Oxford English Dictionary - Old English *slaek*) and is akin to its use in *slaked lime*, or *to slake one's thirst* etc. When sand is made wet, the cohesion that clusters the grains when dry to form dunes is lost and the raised dune topography vanishes, as mobile sand grains form a plane rather than a dune.

When dunes erode the series of flat planes that are formed are usually accompanied by rising water tables. Water can drain downwards rapidly through a loose sandy soil profile, but in planes, increasing wetness with greater depth leads to sand compaction. The resulting density of the sand further impedes drainage seawards and increases the expansion of the area of the slack and with the greater distance from the sea, causes the water table to rise. Figures 3.8-9 also show the boundary up to which trees were planted on the front line of dunes in 1924 as well as the location of the transect which has been used from 1962 to the present day to study water table fluctuations.

In 1962 a transect was marked out across the varied topography of this area (Figure 3.9), starting with the dunes and finishing in a low lying slack by the Powie Burn. This area was then monitored periodically over a period of 24 years from 1963 onwards in order to study the relationship between flooding frequency and and the nature of the vegetation (Crawford et al, 1997). The transect still exists and the water levels are now being monitored electronically to determine if the water relations of these rear slacks have altered over the past 30 years (Sugre in press). In 1979, when

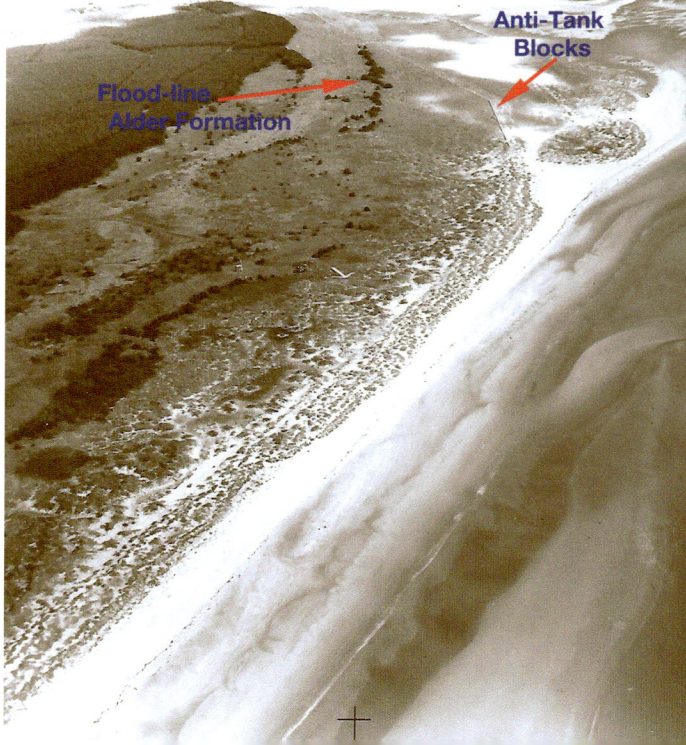

this rear slack was flooded, for a prolonged period, there was also an equally prolonged flooding of the Great Slack to seaward (Figure 3.10-11). Despite the proximity to the sea, the flooding in the Great Slack at this time was usually due to fresh water, with only occasional storm-incursions of salt water.

By the time growth is resumed in spring, monitoring of water extracts from the rooting zone with a salinity meter found only a minimal presence of salt. Evidently, rain can easily remove salt from the upper rooting layers in a sandy soil as there is only a minimal clay content for the retention of salinity.

Permanent wetlands

The ever-present tendency for sand to impede free-drainage has endowed Tentsmuir with a variety of wetlands. Initially, these would have been created in dune slacks formed between the lines of dunes, and would therefore have the same north-south orientation as the dunes. During periods of coastal accretion the front line of dunes advances seawards and the older slacks therefore become increasingly more land-locked and progressively wetter. Those at a distance from the sea are not just prone to flooding in winter, but can maintain high water table levels throughout the year often with the water table remaining typically close to the surface and supporting wetland and even swamp vegetation (see Chapter 4).

Extensive wetlands at Tentsmuir were already evident from the time of the first human presence in the Mesolithic Era (Figure 1.2). Before attempts began to drain Tentsmuir for arable farming some of these wetlands developed into permanent lochs. The most notable and largest of which was *Rossy Myre* as recorded in Bleau's Atlas of 1654 (Figure 2.12).

The tendency for these permanent wetlands to maintain a north-south orientation is due to the fact that they were probably originally dune slacks. The one exception is the wetland that was created artificially in an early attempt at drainage when the Canal Loch was excavated, probably before 1540 (see Chapter 1). This Loch was erroneously named as Candle Loch in Bleau's Atlas of 1654. To this day the site of the former Candle Loch remains an extensive area of wetland in the southern part of the Tentsmuir and is now largely overgrown

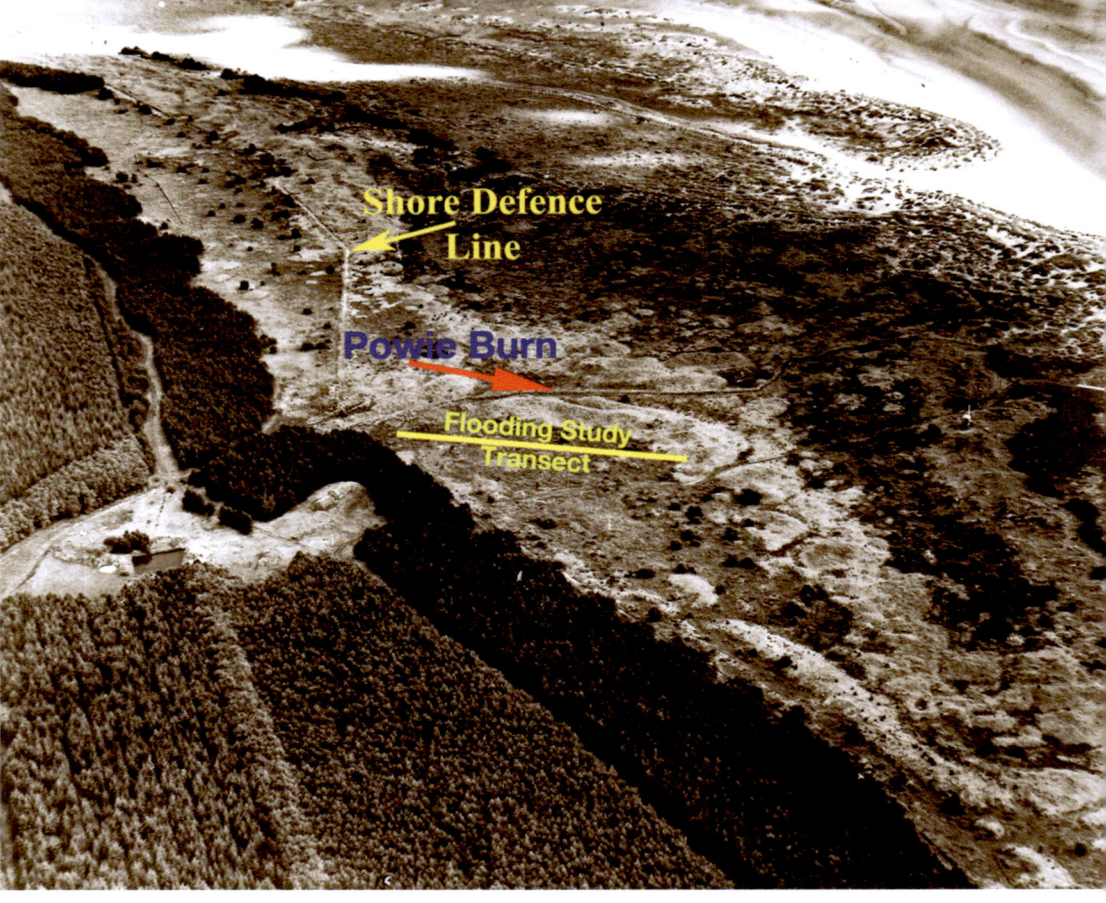

Figure 3.9 Aerial view looking northeast over the Tentsmuir Nature Reserve in 1957 showing dune and slack series. Area selected for a detailed study of flooding duration and depth between 1964-1993 marked in yellow (Photo courtesy of Dr J.K. St. Joseph).

Figure 3.10 Period of winter inundation with the long-term flooding in the study area in the rear slack shown above in Figure 3.9. Note the extent of flooding throughout the greater length of the slack. Photograph taken during a period of prolonged flooding in 1997 (Photo author).

with wetland vegetation and encircled with mixed scrub (Figure 3.12).

Longshore drift

Beaches generally exhibit a phenomenon described as *longshore drift*. This results from the combined forces of erosion and accretion causing the marine sediments to move along the beach in the direction imposed by the local tides.

At Tentsmuir the predominant coastal current is the *St Andrews Bay Current* which runs along the coast of Tentsmuir from south to north. As a result, the sand had a tendency to move north to where it meets the deposits at the Abertay Sands (Figures 2.11-12). From here the sand is blown back by northerly winds onto the front line of dunes at the northern end of the Tentsmuir peninsula. In storm periods this can result in a considerable deposition with dunes heights being increased by over 6 inches in one week (pers. obs.).

The main source of sand that accumulates on the dunes at Tentsmuir Point comes from the floor of St Andrews Bay area, within which sediments have been deposited during late glacial and subsequent times (Browne and Jarvis, 1983).

Figure 3.11 The Great Slack flooded in winter 1979. Although this area is from time to time subjected to salt input from the sea during storms, in general the prolonged flooding is due to fresh water (Photo author).

Figure 3.12 Wetlands with scrub in the southern part of Tentsmuir in the region of the former Canal or Candle Loch (see also Chapter 2 Figure 2.12).

Figure 3.13 Beginning of rapid expansion of dunes and slacks that developed forward of the line of the 1940 concrete blocks. This photograph shows the initial development of a forward patch of embryo dunes in 1973 (Photo courtesy of SNH).

Figure 3.14 Vigorous growth of Lyme Grass (*Leymus arenarius*) on the advancing dunes of Tentsmuir Point in the mid 1970s

The deposits that have accumulated on the Abertay Sands, are made up in part therefore of fluvial elements that have been carried downstream by the river Tay as well as sand of oceanic provenance, coming from the sand circulating in the St Andrews Bay as a whole.

The River Tay is Britain's foremost river in terms of sediment discharge. This is due to its length, and also to the long-standing effects of deforestation, causing extensive erosion in the Scottish Highlands. The discharge has been estimated to average 167 m^3s^{-1} (Jenkins et al., 2005). However, the Highland lochs and the length of the river reduces the eventual fluvial contribution to the Abertay sands.

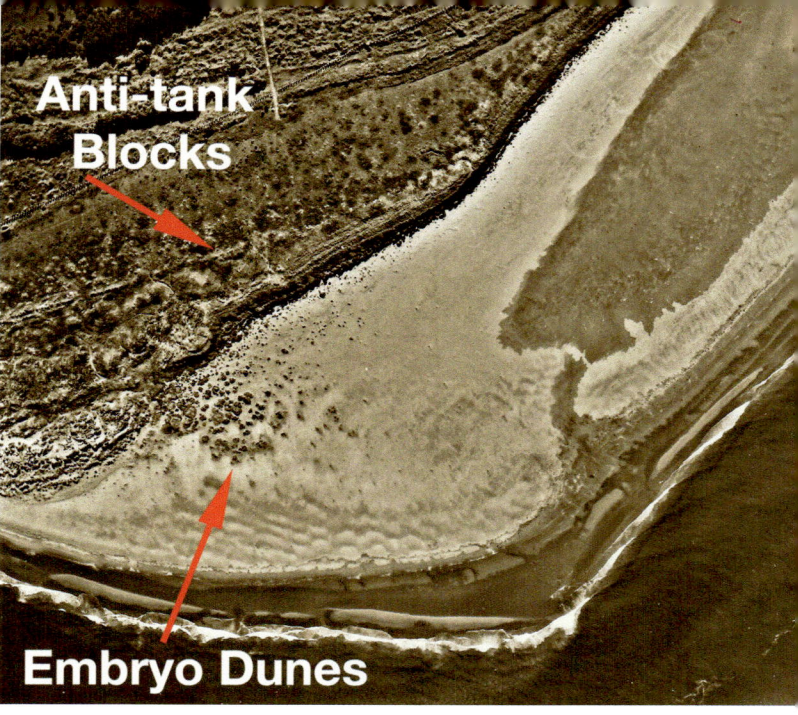

A new insight into the sources of sediment that have brought about the growth of Tentsmuir, has recently been provided by the development of a technique for finger-printing the source of mineral sediments based on the magnetic attributes of the sand grains. A finger-printing study of sediment provenance when been applied to the Tay Estuary (Jenkins et al., 2005). This study showed that the overall contributions of the different sources of sand varied from 78 ± 10%, from marine bottom sediments, 4% from the River Earn, and 18% from the River Tay. Thus, despite the Tay being Britain's foremost river in terms of sediment loading, this research indicates that the fluvial contribution to the sands at the mouth of the River Tay is relatively minor. It appears that this pattern of sediment deposition is consistent with results reported from other temperate estuaries with similar patterns of sand deposition.

Rapid coastal accretion

The closing years of the 20th century saw a period of very rapid coastal accretion at the northern extremity of Tentsmuir. This first became evident around 1973 hasa cluster of embryo dunes appeared to seaward of the main line of dunes at the northern part of the Reserve (Figure 3.13). This was swiftly followed by an expansion of the front line dunes which were consolidated with the rapid spread of Lyme Grass (*Leymus arenarius* – see Figure 3.14).

Throughout the following years from the 1970s to the 1990s there was a remarkable seaward advance of the dune systems at this north east corner of the Peninsula. Certain portions of the shore line regularly recorded a seaward advance of 7 m per annum.

Figure 3.15 Expanse of dunes and slacks that developed forward of the line of the 1940 concrete blocks. This photograph taken in 1977 shows the extent of the coastal accretion towards the end of the 20th century

Sand and water

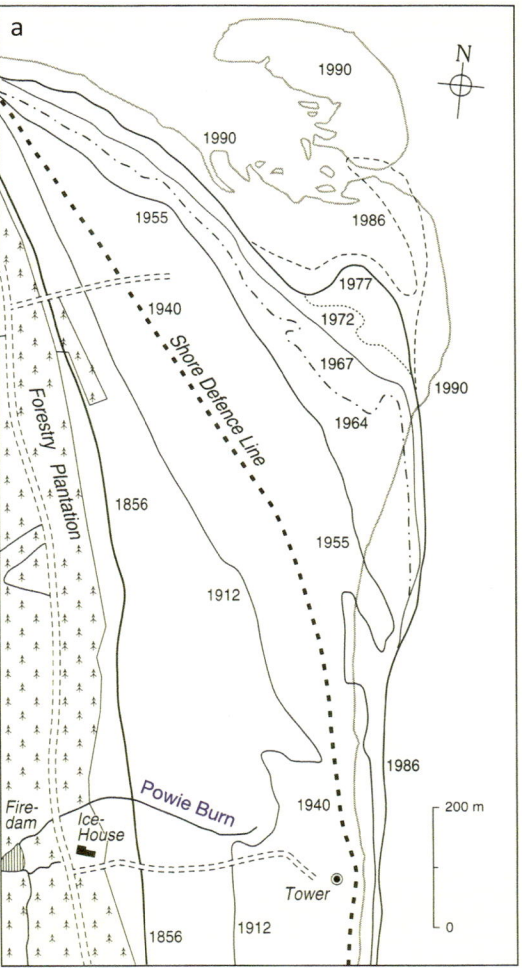

Figure 3.16 Map and photograph of coastal change at Tentsmuir (a) accretion 1856-1990 (Crawford, 1996); (b) erosion as recorded by satellite image (Google Earth 2015).

The rapid expansion of the fore-dunes was greatly promoted, and for a while stabilized, by the vigorous growth of Lyme Grass (*Leymus arenarius*). This species grows best on the seaward face of the front line of dunes facing the sea. The root systems are shallow and extend horizontally which favours their ability to colonize fresh deposits of sand. This growth habit in Lyme Grass is in sharp contrast to Marram Grass (*Ammophila arenaria*) which has roots that descend vertically through the sand dune. When the two sand-dune grasses are present the Lyme grass is normally dominant up to the dune crest on the seaward side of the dune, with Marram then becoming the main grass species on the landward side of the dunes.

The vast expansion that has taken place at Tentsmuir with the growth of the new dunes and slacks since 1940 is shown in Figure 3.15.

Coastal accretion time-scale

Eighteen thousand years ago Tentsmuir and its sands did not exist. The carbon dating of the first occupation of the Mesolithic site at Morton provides an approximate date-marker for the shoreline at 10,000 years B.P. Calculating from the present position of the shoreline in relation to the location of the Mesolithic site (see Chapter 1), it is possible to obtain a long-term figure for the expansion of the dunes and slacks that developed forward of the anti-tank blocks. Over this 10,000 year post-glacial history, the Tentsmuir Peninsula would have undergone retreats as well as advances. However, when averaged for the overall rate of advance seawards of the coastline, the rate of growth at the more northern part appears to have averaged 0.5 m per annum for the last 10 millennia (see Chapter 1).

Taking a later starting date for assessing coastal advance from the stone marker for a salmon fishing boundary erected in 1794 (map ref. NO 499 272) to the position of the coastal defence blocks (Figure 3.17) put in place in 1940 is a distance of 650 m which gives an average accretion rate of more than 4.5 m.p.a. over a period of of 146 years. From the 1940 blocks to the furthest extension of the dunes in 1990 was approximately 350 m. This gives an accretion rate of 7 m.p.a.

Another recent study of this same area (Duck, 2011) in relation to coastal accretion has estimated the growth rate at a little less than 5 m.p.a. with the comment *that this, if*

Figure 3.17 Anti-tank blocks on the lower shore with many sinking below the waves. Photograph taken in August 2005.

Figure 3.18 Aerial view of the northern part of Tentsmuir bordering the southern shore of the Tay Estuary in 2012. A small southern portion of the Abertay Sands is visible to the bottom left (Photo courtesy of Gordon Pickthall).

not the most rapid, this is certainly one of the most sustained rates of growth on one of the most swiftly accreting parts of the British coast.

In the same general area McManus and Wal (1996) estimated that the linear growth seaward of the vegetated area between 1812 and 1990 on a line perpendicular to the coast was for a movement of approximately 500m eastward, giving an average growth of approximately 2.8m annually eastwards over a period of 178 years.

The actual date for the maximum extension of the dune system before erosion set in at this region is not certain. Estimates vary depending on a location and whether growth is measured as the total new area added, or merely the seaward linear advance of the sand dunes. Whatever method is used, it would appear that the maximum extension of the Tentsmuir Point Peninsula took place between 1990 and the end of the millennium. Despite a recent period of very active erosion (see below), the Tentsmuir Point Nature Reserve is still considerably larger that when it was created in 1954.

It is difficult to be certain as to the source of the sand that brought about the recent advances of the coastline. However, it has been suggested that the entrapment of the sand was from erosion that took place near the Kinshaldy region of the coast. Using a combination of aerial photography and contoured maps of the Reserve it has been estimated that the accumulation volume of sediment accumulated at Tentsmuir Point between 1978 and 1990 was $33 \times 10^3 \, m^3$.

Erosion

One of the most astonishing aspects of change in the Tentsmuir landscape in recent decades is the speed

Figure 3.19 Observation tower erected on the front line of dunes for observation of RAF target practice during the 1939-45 war. It collapsed on the 29th December 1998.

with which rapid coastal accretion has been replaced by even more rapid erosion. This dramatic reversal of the relationship between sand and water is recorded in concrete in the history of the coastal anti-tank blocks. Most of the blocks that were originally placed on the high tide mark were, by the latter part of the 20th century to be found either well inland or else invisible, as they were completely buried in sand. Now they are either well down on the shore, or else sunk below the waves. (Figures 3.17-20).

Similarly, the volume of sand eroded from the more southern margins was estimated to be 46×10^4 m^3 (McManus and Wal, 1996). This trend has recently been partially reversed as the Kinshaldy region appears to be gaining from the erosion of the northern extremity of the Peninsula (see page 67).

A new Great Slack that had begun to be formed at the far north-east end of the Reserve has now been completely removed and only a few small sand dunes remain. This trend for erosion at at the northern extremity of the Peninsula has recently been reversed at the more southern end in the Kinshaldy region which appears to be gaining from the erosion of the northern extremity of the Peninsula (see page 62). Seawards of the Ice House the land has retreated back to approximately where it was in 1912. Not only have the dunes been completely removed, but also well-known wartime landmarks, such as the observation tower for monitoring bombing practice and the concrete arrow for directing bombing practice flights have now all disappeared beneath the waves. (Figure 3.19).

Sand and water   45

a

b

Figure 3.20 The fate of the alder flood line association. (a) Alders at edge of beach 2008: (b) Alder on beach October 2010 (c) The Green Hut survives where it was originally placed over half a century ago to observe terns nesting on the now vanished foreshore. The hut and trees have now gone (Photograph taken 2014).

c

In 1940 the tower was 150 m from the dune edge. By 1997 the tower was less than 9 m from the sea and on the following year the tower fell onto the shore on the 29th December. Even more destructive was the obliteration of most of the well established *Alder flood-association* trees (Figure 3.20 a-c). The average linear rate of coastal retreat from 1990-2018 for this part of the Reserve has averaged 10.2 m per annum.

It has often been noticed that when sand is removed from one place it is usually replaced in another. A figure that was often quoted in the 20th century was that Scotland lost and gained about 300 acres of shoreland p.a. However, this comforting statistic may no longer be valid in the face of rising sea levels. Surprisingly, despite the loss of terrain at the more northern part of Tentsmuir, principally in the Nature Reserve, Tentsmuir is at present gaining ground further south in the region at Kinshaldy.

The question has therefore arisen as to how this could have happened and whether it is the sand lost from the north that is supplying the growth of dunes to the south. Given the presumed south to north direction of the St Andrews Bay Current (Figure 3.24) this would not appear to be a direct

transfer by the longshore drift unless it has had a change of direction. Alternatively, the sand might be transferred by a wider circulation route moving the sediments in a clockwise direction out into St Andrews Bay and then returning the sand at the southern end (see Figures 3.21-22 for recent changes over past millennia). At present attempting to account for the source of this newly accreted land at the region of the Kinshaldy beach is largely speculative.

Figure 3.21 Accretion that has taken place this century at the more southern region of Tentsmuir at Kinshaldy where embryo dunes have advanced significantly seawards in the past decade (Photo. from Google Maps 2015)

Long-term history of the Tentsmuir coast

The presence of the Abertay Sands to the north of Tentsmuir creates the expectation that this proximal source of sand contributes to the coastal accretion of the coastline. However, historical maps that have been produced from borings made into the local coastal deposits, rather than direct observation of changes of the the coastline, tend to provide a different view of the ancient outline of the Tentsmuir Peninsula. Already in the 1970s

Figure 3.22 View facing north showing the newly accreted sand and embryo dunes seen above in the satellite image (Figure 3.21) and the recent extensive deposits of sand with the newly developed embryo dunes.

Figure 3.23 A sand dune island at Tentsmuir Point in August 2015. Such islands may have been a prominent feature in the past (see Figure 3.25) and may become a feature once again as a result of storms and rising sea levels.

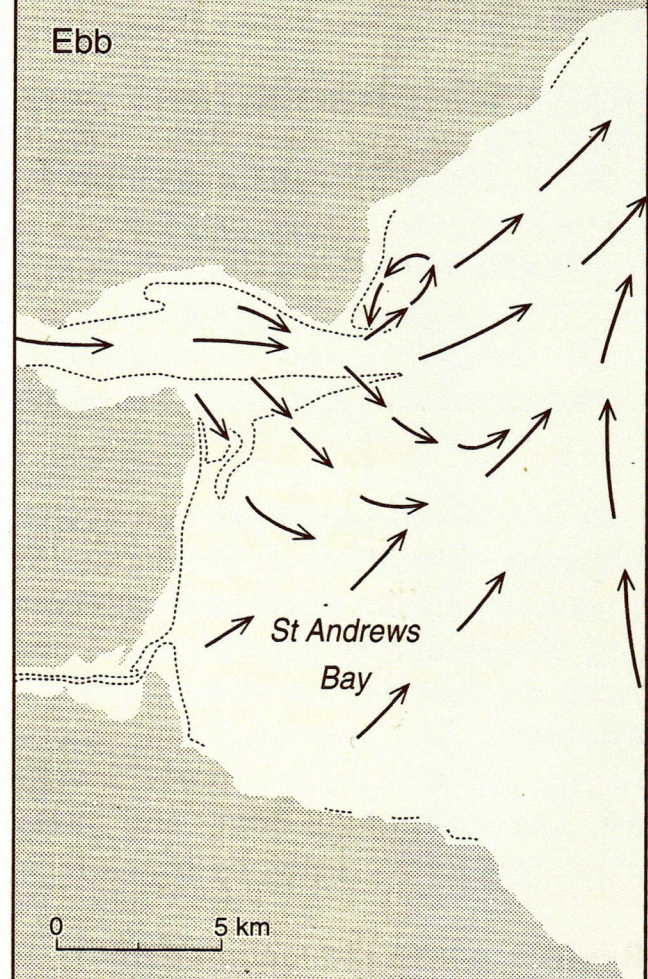

Figure 3.24 Tidal circulation patterns in St Andrews Bay (Reproduced with permission from McManus and Abhilsha Wal 1996.

48   Tentsmuir: Ten Thousand Years of Environmental History

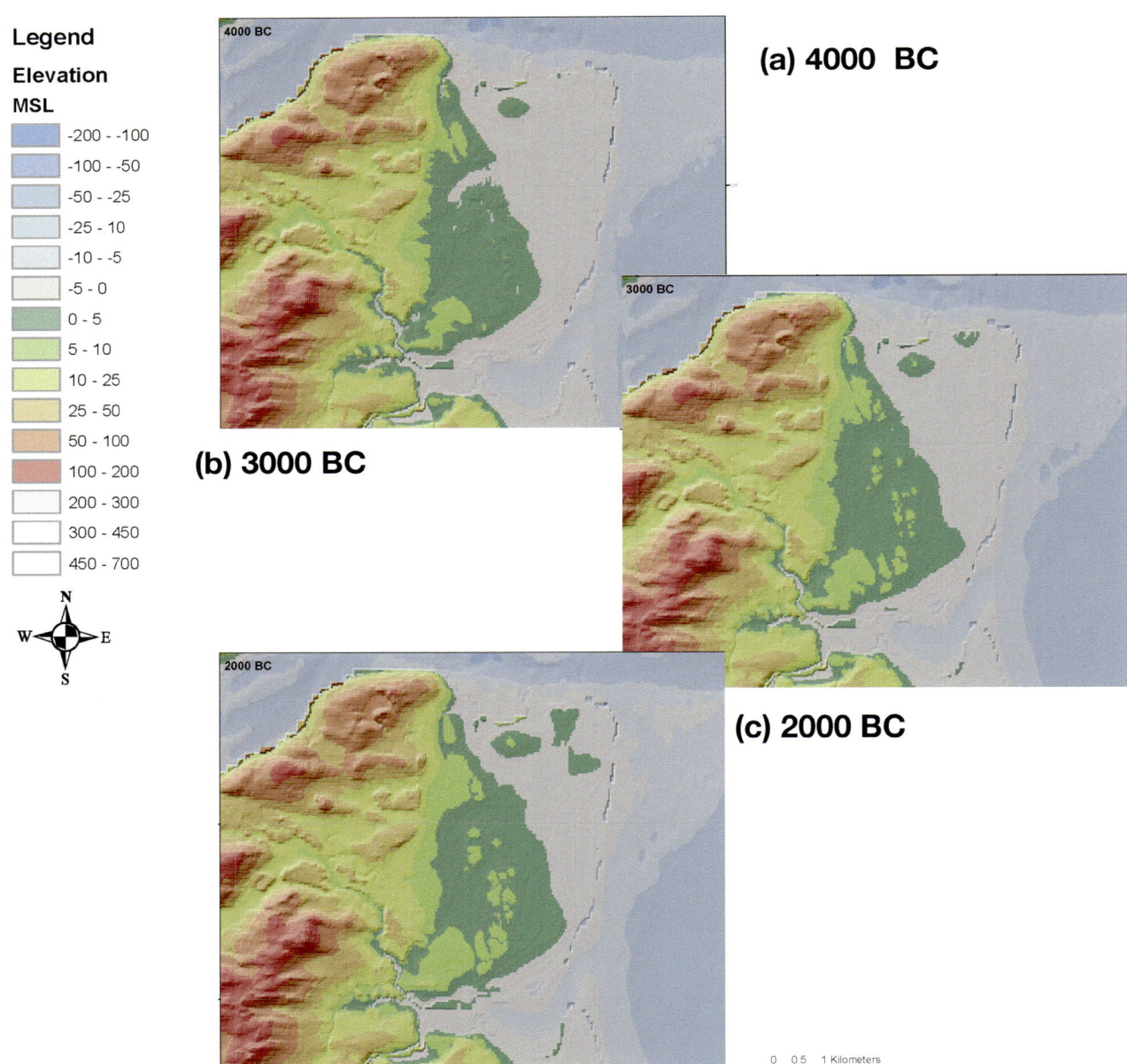

Figure 3.25 Historical model of the coastline history of Tentsmuir based on tidal movements and the geometry of the adjacent sea floor at Tentsmuir from 4000 to 2000 BC as assessed from coastal borings of adjacent sea floor (Images courtesy of Professor R.J. Bradley and Dr. F Sturt).

a map produced in relation to the archaeological study of the Mesolithic presence at Tentsmuir *c.* 8000 BC showed the more southern region of the Tentsmuir Peninsula as having a greater seaward prominence than the region to the north (Coles, 1964; Coles, 1971).

A recent study using a model that combines calculations of sea level change, shoreline displacement, terrestrial topography, submarine contours and borehole data (where these are available) has been used to construct coastal maps at 500 year intervals from palaeogeographic models for Britain, Ireland and the North West French coast from 11000 cal. BP to present (Sturt et al., 2013).

This model has been used to obtain representations of possible coastline changes that may have taken place at Tentsmuir over the past 6000 years. Figure 3.25 indicates the presence of small coastal islands in the northern part and a coastline that has advanced seawards particularly at the southern end of the Peninsula, where it has moved seawards to a greater extent then at the north. If this has been the more stable long-term configuration of the Tentsmuir coastline in the past then the present growth shown in the southern region near the Kinshaldy Burn might merely be a return to the more habitual shape of the Tentsmuir Peninsula. Such a possibility is a salutary reminder of inherent uncertainties in human attempts to predict the outcome of the action of natural forces in controlling the actions of sand and water in shaping the landscape.

The future survival of Tentsmuir

Tentsmuir has a soft coast. It has no hard rock promontories, and the only cliffs are the ancient sea cliffs at Craigie Hill which formed the coastline at the end of the Pleistocene glaciations. In common with other low-lying areas in Britain, it may therefore face an uncertain future. Britain has many areas where the coastline is fragile (Duck, 2011). The Fens, and particularly the region of the Wash, are areas where substantial loss of land to the sea may take place, not just because the sea level is rising, but because the drainage of peat and the consequent shrinkage of the organic soil profile, plus wind erosion of the dry, crumbly, peat is causing the level of the sea in relation to the coast to rise (Pryor, 2004). Similarly, at Tentsmuir, where much of the land is only 1m above sea level, erosion of this soft coastline by an increase in the frequency of storms may cause a substantial retreat even without a significant rise in sea level.

The long-term future of Tentsmuir will therefore depend on the relationship between sand and water. The uncertain question, of whether tidal currents and wind, will add sufficient sand to the Tentsmuir shore to save it from erosion is far from certain.

It is always a possibility that rising sea levels will cause increased disturbance to the vegetation and sand dunes which might then migrate inland and engulf the forest.

Chapter Four

Tentsmuir's dunes – a changing landscape

Dune structure and stability

The greatest change to take place in the past 100 years on Tentsmuir's dunes was the establishment in 1924 of a substantial tree plantation by the Forestry Commission. Having acquired the greater part of the Tentsmuir Peninsula, the foresters proceeded to plant trees up to just behind the front line of dunes (Figures 4.1-2). This forest, as it stands today, consists mainly of Scots and Corsican Pine (*Pinus sylvestris, P. nigra*), and Norway Spruce (*Picea abies*). Now that it has matured, the forest has totally altered the coastal environment of the greater part of the Tentsmuir Peninsula from what it was in the past. Throughout the greater part of the 19th century, the non-agricultural northern portion of Tentsmuir, had a robust but precarious existence as a grouse moor and duck shoot (see Chapter 6). About 1870 a bad fire, followed by a very severe gale on the 17th November 1873, had left an area of drifting sand, creating a landscape with a desert-like appearance – not unlike the Culbin sands in their earlier treeless condition but on a smaller scale (Ross, 1993).

Grouse were first introduced to Tentsmuir in 1876 and then again in 1894 (see Chapter 1). The population remained low but was nevertheless healthy, as regular flooding with salt water protected the birds from the condition often referred at that time as *'grouse disease'* which caused regular population crashes. This disease was then prevalent in Scotland, and a canal had purposely been cut to facilitate the salination of

Figure 4.1 Satellite image of Tentsmuir showing the location of the Forestry Commission planting blocks and the proximity of the forestry plantations to the shore (Google Earth 2018).

Figure 4.2 Details of Tentsmuir Point showing the proximity of the forest planting to the sea. Note the 1940 anti-tank blocks shown by the red arrow indicating the location of the the coastline in 1940. (Photo -SNH)

the slacks which helped to reduce the incidence of the disease at Tentsmuir. Consequently, fresh and salt water marshes, together with pools of fresh water existed in close juxtaposition to areas of waste-land and shifting sands. The whole area was populated with rabbits to such an extent that it justified the employment of two rabbit trappers (Smith, 1948).

Dr John Berry (1907-2002) (see frontispiece) who arranged the purchase of the Nature Reserve from the Forestry Commission, and whose father had in the past rented part of Tentsmuir for shooting, knew this area intimately from a very early age. His fifth birthday was celebrated at Morton Lochs in 1912 with the Misses Baxter and Rintoul, the distinguished Scottish ornithologists (see Chapter 6).

In 1995 Dr Berry recorded on tape (for the author) his boyhood memories of the pre-afforestation state of Tentsmuir as it was in the early years of the 20th century. His predominant recollection, was of a coastal moorland with much blown sand and many mobile dunes, creating an overall appearance resembling a desert wasteland. He remembered the region particularly as having very many high dunes, with some over 50 feet high. A number of these high dunes had probably formed over buried forest trees. There was also an extensive barren moorland interspersed

Figure 4.3 Colonizing rhizomes of Sand Couch Grass (*Elytrigia juncea*) advancing onto fresh sands at Tentsmuir in 2015 in the area of recent dune advance at Kinshaldy (Photo author).

Figure 4.4 Sand Couch Grass (*Elytrigia juncea*) forming embryo dunes on the shore at Tentsmuir Point (Photo author).

Tentsmuir's dunes – a changing landscape

Figure 4.5 Lyme-grass (*Leymus arenarius*) building up a significant presence and anchoring newly accreted sand near Kinshaldy 2015 (Photo author).

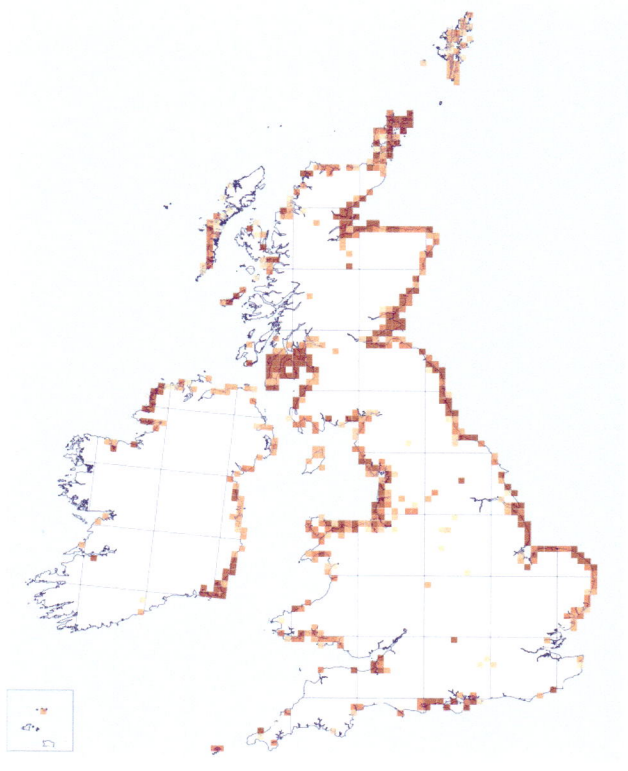

Figure 4.6 Distribution in Britain and Ireland of Lyme-grass (*Leymus arenarius*). (Reproduced with permission from Preston CD, Pearman A, Dines TD. New Atlas of the British and Irish flora: Oxford University Press; 2002.)

with remnants of forest. During periods of high tides, salt water penetrated well into the moor where there was a large brackish pool. The influence of high salt concentrations, drought, erosion and flooding, provided a shifting and unstable landscape with much bare ground, generally evoking an impression of great physical fragility (see also Chapter 9).

Physical versus biological fragility

Many plant species, and particularly those that comprise coastal floras, are adapted to growing in physically disturbed sites. Consequently, a degree of disturbance can be necessary to provide opportunities for their regeneration and the maintenance of coastal habitat diversity. As a result, many dune systems may look fractionated and damaged. Nevertheless, they still posses an inbuilt capacity for recovery, provided there is the possibility of the flora being restored with propagules from neighbouring undamaged sites.

The coastal location of dune systems such as Tentsmuir, ensures that there is usually access to propagules that are either washed up, or blown along the foreshore from similar habitats up and down the coast. Many dune sedges and grasses have extensive rhizome systems which become detached in storms and then root elsewhere. Disturbance in such situations aids dispersal and regeneration.

In the latter part of the 20th century the growth of the dune system at the northern part of Tentsmuir was so rapid (see Chapter 2) that it tended to favour the development of grass-dominated vegetation to the detriment of other flowering plants. It was in fact, the physical fragility of the dunes and slacks that created their florisitic heterogeneity which had been the delight of naturalists and botanists in the early years of the 20th century (Crapper, 1939; Wilson, 1910).

Sand dune grasses

In common with many northern European sand-dune systems, Tentsmuir is dependent on grasses for the creation and stabilisation of the coastal sand deposits. A dynamic interaction can be seen between the three major grass species that form the basic vegetation cover of the front line of dunes. The first of these grass species to become established on the foreshore is usually Sand Couch Grass (*Elytrigia juncea* Figures 4.3-4).

When Sand Couch Grass first appears it can often be seen advancing across the sand in straight lines with its rhizomes situated just below the surface (Figure 4.3) This creates the initial embryo dunes that usually develop seawards of the main line of dunes. Sand couch grass can spread readily and in doing so, traps the initial accumulations of sand on the shore.

Figure 4.7 *Marram (Ammophila arenaria*) establishing itself on the top of an embryo dune. Note the presence to seaward of Sand Couch Grass (*Elytrigia juncea*) that probably initially formed this forward dune (Photo author).

Despite its capacity to thrive on a very mobile substrate, Sand Couch Grass is very sensitive in relation to the amount of sand burial that it can withstand. Experiments have shown that although both newly-germinated seedlings and fragments of rhizomes can emerge from a burial of 127 mm of sand, nevertheless they will fail to do so if the burial exceeds 178 mm (Harris and Davy, 1986).

Due to the burial limits of the pioneering dune grasses, embryo dunes are limited in the height to which they can grow until some other more burial-tolerant grass species arrive and take over the anchoring of the sand. In many locations in Northern Europe, and also at Tentsmuir, Lyme-grass (*Leymus arenaria* Figures 4.5-6) is a frequent colonizer of the sea facing slopes of the frontal dunes.

At Tentsmuir this species was largely responsible for anchoring the large quantities of sand that were driven on shore resulting in the rapid accretion that took place in the latter part of the 20th century. To landward of the Lyme Grass, the commonest sand dune grass Marram (*Ammophila arenaria*) becomes dominant (Figure 4.7).

At the end of the 20th century, the major front line of dunes were low gradient beaches which varied in their seaward extension and could be up to 400m wide. As a result, the beach system at Tentsmuir responded rapidly to changes in wave action and modification to tidal conditions. The Abertay Sands which extend eastwards for 6 km beyond Tentsmuir Point, provided extensive sources of sand during periods of onshore winds (McManus and Alizai, 1987).

The pattern of dune growth has changed since the end of the last millennium. The Lyme Grass is now entirely removed by erosion from near Tentsmuir Point. Instead, extensive deposits of sand are now being consolidated by various sand dune grasses further south at Kinshaldy as discussed in Chapter 3 (Figures 3.22-3). Paradoxically, loss of vigour in sand-dune grasses, in the seemingly favourable environment of well-established dunes, has long intrigued ecologists (Moore, 1996). In particular, *Ammophila decline* is a phenonmenon that has attracted attention in the north and mid-Atlantic coasts of the United States, where the American *Marram* species (*Ammophila brevigulata*) is showing extensive die-back.

Marram Grass (*Ammopila arenaria* Figure 4.7) is the most common grass found generally in the front line of dunes. It is widespread in Europe as is the closely related species *Ammophila brevigulata* in North America. These two species are widespread as effective dune-stabilizers and can maintain their presence even with vertical sand accretions rates of up to one metre per annum. However, despite their initial pioneering success they frequently lose vigour as the sand levels stabilize.

Fresh burial of the stems of this plant have always appeared to be necessary for the continued vigour of Marram Grass and a variety of explanations, including the need for new rhizome bud development, and the adverse effects of soil compaction have all been discussed at various times. Both European and American Marram species have been shown to benefit from burial by sterile sand, possibly as an escape from pathogens and in particular, nematode attack. It has been found however, that it is not just the sterile soil *per se* which provides the escape from nematode attack, but the facility this sand provides for the development of fungal associations (mycorrhizal connections) which give the plants the vigour necessary to combat nematode infections (Little and Maun, 1996). Thus, periodic destruction of Marram covered sand dunes may be a necessary feature for their long-term rejuvenation and survival.

Plant geography at Tentsmuir

An important factor, already mentioned in Chapter 1, is the favoured location that Tentsmuir has climatically for plant growth. This can also be said about its latitudinal position which allows for migration both north and south along the shoreline. It is therefore not surprising that Tentsmuir provides habitats for plants that have generally either a more northern distribution than expected, yet at their northerly locations also provide habitats suitable for species with more southerly distributions.

The Baltic Rush (*Juncus balticus*) is a species with a northern distribution that used to be abundant at Tentsmuir (Figures 4.8-9). For many years this species was common on the edge of the Great Slack to seaward of the concrete blocks. This is still the most southerly location of this species on the east coast of Scotland. However, the Baltic Rush appears to have been retreating northwards in recent years, nevertheless a sizeable population still remains at Tentsmuir. Internationally, botanists consider the Baltic Rush as a sub-species of the circum-polar *Juncus arcticus*.

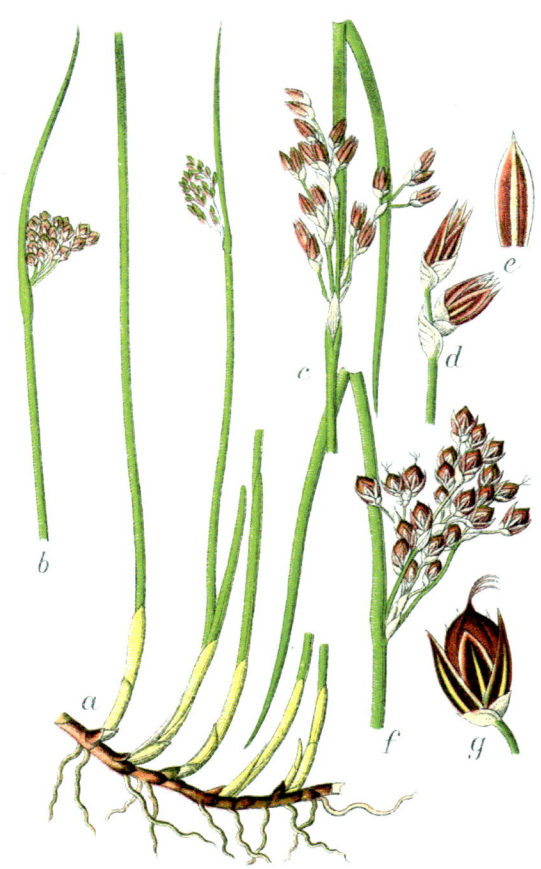

Figure 4.8 The Baltic Rush (*Juncus balticus*). Figure from *Deutschlands Flora in Abbildungen* (1885) Johann Georg Sturm (Painter: Jacob Sturm)

The Coral Root Orchid (*Corallorhiza trifida* – Figures 4.10-11) is an example of a northern species which was once well-known at Tentsmuir. Elias Crapper was a local botanist from Tayport who made a detailed study of Tentsmuir in the first half of the 20th century and put together in 1940 a Flora of Tentsmuir which sadly has never been published but is now lodged in the archives of St Andrews University (Crapper, 1939). He visited Tentsmuir regularly over may years with

Tentsmuir's dunes – a changing landscape   57

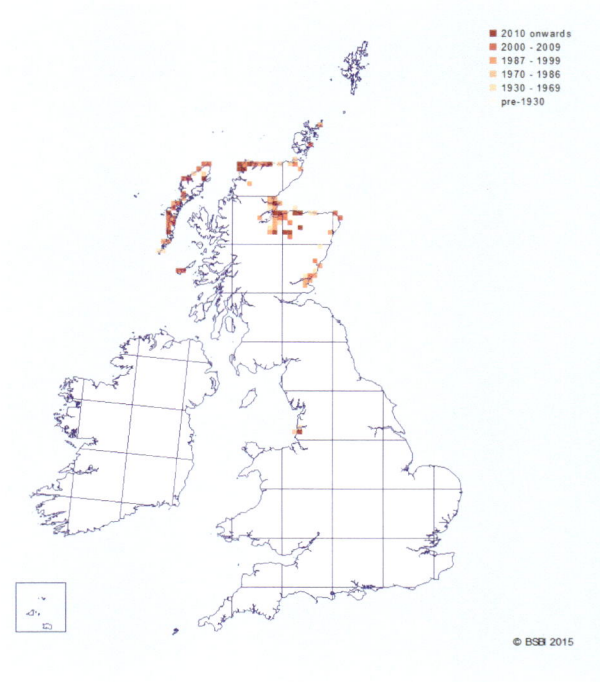

Figure 4.9 Distribution in Britain and Ireland of the Baltic Rush. Note extinct site on the Lancashire Coast. (Reproduced with permission from Preston et al., New Atlas of the British and Irish flora. (Reproduced – Atlas of the British Flora, Oxford University Press, 2002).

Figure 4.10 Coralroot Orchid (*Corallorhiza trifida*) - a saprophytic herb that inhabits shaded damp Willow and Alder carr as well as freshwater dune slacks with *Salix repens* (Photo T. Cunningham).

generally detrimental to many of the flowering plants species and not just the Coral Root Orchid.

The Coral Root Orchid (Figures 4.10-11) is a species that occurs typically in Alder and Willow carr, as well as in dune slacks where it is commonly found in the interface zone between the edge of the tree cover and open ground. At Tentsmuir in the mid-20th century this was one of the foremost sites in Britain for the occurrence of this species (Figure 4.11).

his friend the well known local naturalist and artist Len Fullerton, many of whose paintings are now also lodged in the archives of St Andrews University and Crapper gives a vivid account of the distribution of Coral Root Orchid, which he describes as a species *'that was always regarded as Tentsmuir's most interesting plant, and many botanists have vistied the moor to find it, and not always successfully.'*

The truth of these remarks is now all too evident. The proximity and size of the forest and its enormous transpiration potential, together with improved drainage for the trees, has reduced the level of the water table generally at Tentsmuir. This is now proving

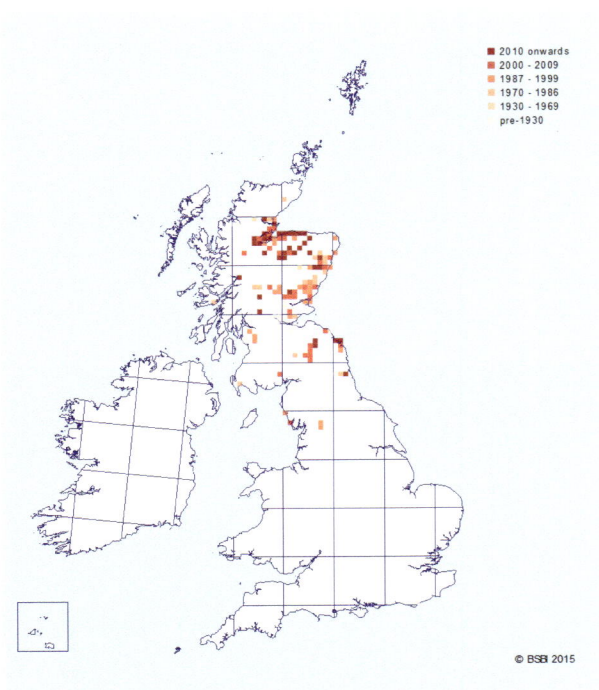

Recently, a painting of the Twin Flower by Len Fullerton has been found in a collection of his paintings donated by his family to the archives of St Andrews University. It would appear likely that this was painted when the species was found in the 1940s, most probably in the vicinity of the Ice House.

Given the rarity of this species in recent years (Figure 4.13) it is very fortunate that the Twin Flower has been once again recorded and confirmed at Tentsmuir by the Botanical Society of the British Isles. Overall, this species appears to be currently retreating from its more southern locations. Whether or not this is due directly to climatic

Figure 4.11 Distribution of the Coral Root Orchid (*Corallorhiza trifida*). (Reproduced with permission from Preston et al., New Atlas of the British and Irish flora. Oxford University Press, 2002).

The recent erosion that has taken place at Tentsmuir has also removed much Alder Carr that was once the habitat of the Coral Root Orchid. There has also been a general decline elsewhere in the occurrence of the species as is evident in the maps of its past occurrence in the British Isles. (Figure 4.11).

Another rare species recorded by Crapper was the Twin Flower (*Linnaea borealis*) Recently, there here has been a rediscovery of this species at Tentsmuir and elsewhere in the east of Scotland. This plant was already recorded there by Elias Crapper in the 1940s and painted by Len Fullerton (Figures 4.12-13).

Figure 4.12 Specimen recorded by Crapper (1940) of Twin Flower (*Linnaea borealis*) found growing near the Ice House and painted by Len Fullerton. (St Andrews University Archives).

Tentsmuir's dunes – a changing landscape

warming, or from other causes such as habitat alteration which has also taken place at Tentsmuir due to the removal of Alder and Birch is not clear. There has also been a national decline in the occurrence of this species as is evident from the maps of its past and present occurrence (Figure 4.13)

To the rear of the front line of dunes, where less fresh sand is available, the grasses lose their dominance. and the dunes become liable to blow outs and other forms of erosion. This gradual erosion of the dunes increases the area covered by dune slacks. At Tentsmuir it was this sheltered landward facing side of the dunes that used to carry a rich flora of flowering plants among which both the Common and Coastal Centaury (*Centaurium erythraea*

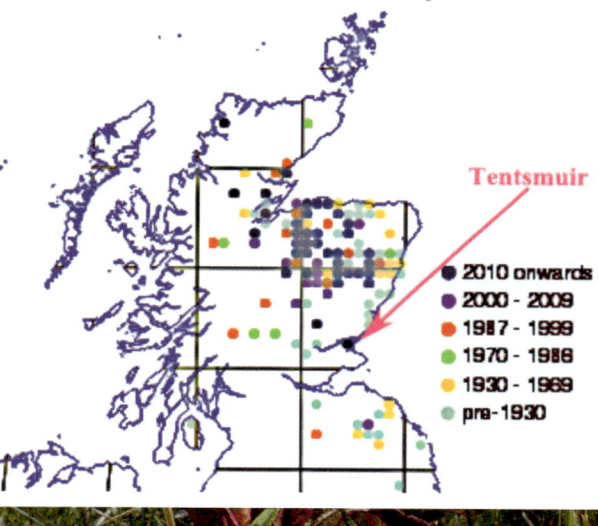

Figure 4.13 Map showing historical records for the occurrences of *Linnaea borealis* in Scotland. Note that Tentsmuir is the most southerly known current record of this species. (Data from the records of The Botanical Society of the British Isles).

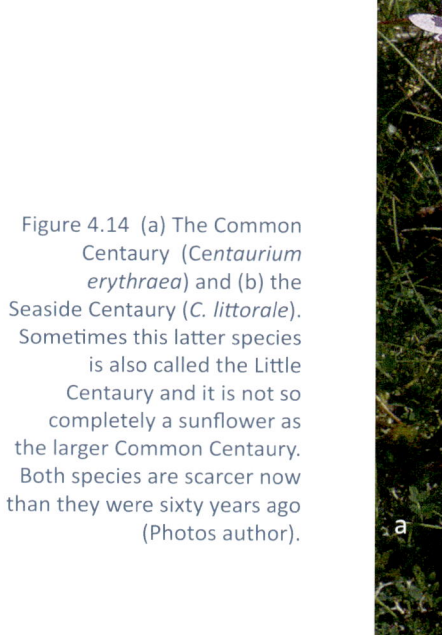

Figure 4.14 (a) The Common Centaury (*Centaurium erythraea*) and (b) the Seaside Centaury (*C. littorale*). Sometimes this latter species is also called the Little Centaury and it is not so completely a sunflower as the larger Common Centaury. Both species are scarcer now than they were sixty years ago (Photos author).

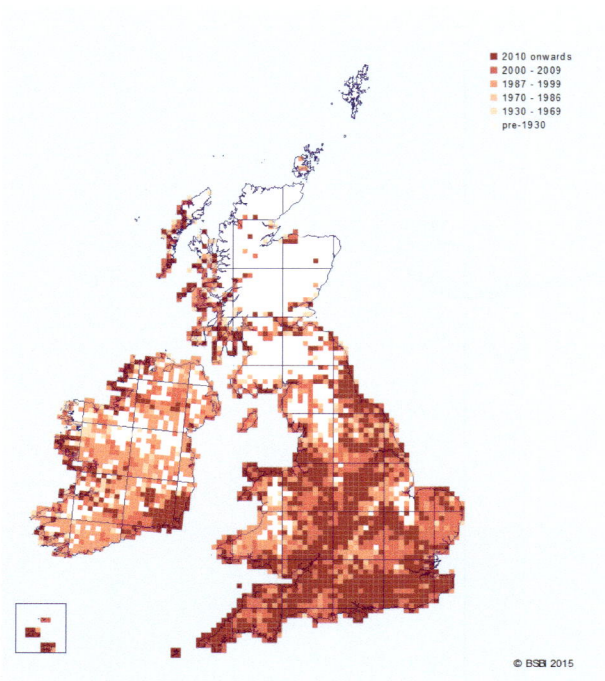 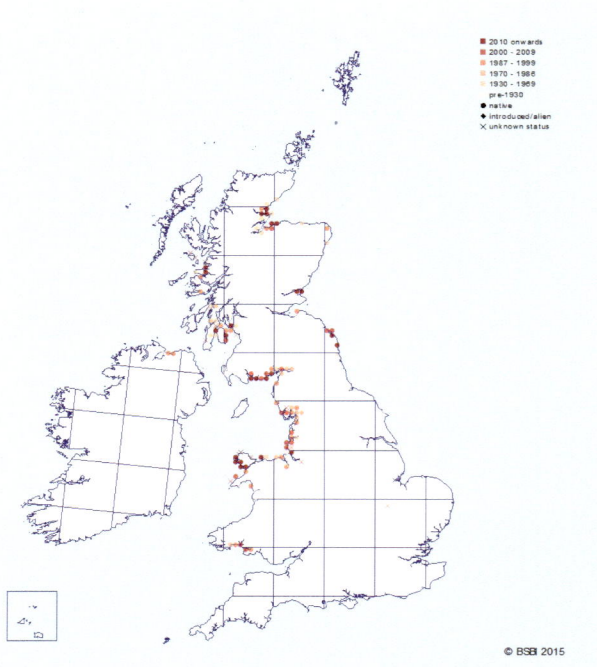

Figure 4.15 British distribution in Britain and Ireland of (a) the Common Centaury (*Centaurium erythraea*), (b) the Seaside Centaury (*C. littorale*). (Reproduced with permission from Preston CD, Pearman A, Dines TD. New Atlas of the British and Irish flora. Oxford: Oxford University Press, 2002).

and *C. littorale*) were a feature of the summer flowering season. Sixty years ago the landward sheltered side of the fore-dunes were favourite locations for observing these two Centaury species but they have now become less common at Tentsmuir (Figures 4.14-15).

The outstanding part of Tentsmuir mentioned by Crapper for floral displays in Summer is the region near the Ice House. Here he particularly noted abundant displays of the Grass of Parnassus (*Parnassia palustris* Figure 4.16) which he described as *'without doubt the aristocrat of all the flowers to be found on Tentsmuir, for to see it in a good season (and a good seasosn means a wet one), thickly covering over an acre, is something to be remembered.'* He also notes that *'here again we have another of the outstanding flowers of Tentsmuir that is strictly confined to one part of the moor. The vicinity of the Ice House supplies the home, and a very little search east and north east of that building will produce the plant in plenty.'* Fortunately, there are now signs of recovery for this species.

A field party of botanists recently rediscovered the Field Gentian, *(Gentianella campestris)* growing in the Great Slack. They were particularly pleased to be able to list this species again in the Tentsmuir flora. It had not been seen for several years even although it was recorded at Tentsmuir in the 2002 Atlas of the British Flora (Figure 4.17).

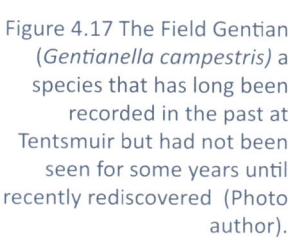

Figure 4.16 Grass of Parnassus (*Parnassia palustris*) growing at the edge of the Great Slack Tentsmuir 2015) (Photo author).

Figure 4.17 The Field Gentian (*Gentianella campestris*) a species that has long been recorded in the past at Tentsmuir but had not been seen for some years until recently rediscovered (Photo author).

Figure 4.18 Invasion of Fireweed (*Chamerion angustifolium*) onto the land-facing side of the Fore-dunes at Tentsmuir Point (Photo author).

Curiously, this handsome plant is described by Crapper (1939) as '*a disapointment,*' as '*being a member of the brilliant family. With the Gentians, one is led to expect it to be a flower of some beauty, but sad to relate it posseses no particular beauty or charm of any kind.*' Nevertheless everyone was pleased to be able to list this previously despised, but apparently attractive species flourishing again in the Tentsmuir flora.

Similarly, the recording once again of a colony of the Twin Flower (*Linnaea borealis*) is a welcome recent addition to the current flora, especially as it is has recently been reappearing at various sites in eastern Scotland (Stace, 2010).

Sadly, in recent years many of the floristic delights of the Tentsmuir dune flora so evident in the past, have become obscured and suppressed by the spread of Rosebay Willow Herb, (also called Fireweed, (*Chamerion angustifolium*). This species has invaded the dunes and now dominates large areas

of the more landward facing dune slopes (Figure 4.18). In the 18th century Rosebay Willow Herb was considered to be rare in Britain. However, with the arrival of railways and their cinder tracks, it spread widely throughout the whole of the British mainland. It also increased its presence on bombed sites during the Second World War.

Fireweed has long had a circumpolar distribution in the Boreal Forest where it probably achieved its ecological success through an ability to germinate, produce initial roots and initiate growth in the low nutrient environment of a forest floor, due to a well-developed capacity for nutrient capture (Pinno et al., 2014).

There are two forms of Fireweed, one lowland and the other montane. The lowland form of this species was probably introduced from Canada and has achieved its current presence in Britain over the past 150 years. It is now invading sand dunes where its presence has been facilitated by the large numbers of rabbits in many coastal areas before their numbers were reduced with onset of the myxomatosis disease in the early 1950s. An increased content of nitrogen in rainfall in recent times may also be a significant factor in contributing to eutrophication and the success of Rosebay Willow Herb as in the in the past it would have been nitrogen-limited in sand dunes.

When Fireweed reaches coastal sites, as it has done in recent years at Tentsmuir, it can dominate the nutrient-poor sand dunes (Figure 4.18). It has not however completely obliterated species such as the Common Centaury, although it does over-top these smaller plants and diligent search is required to find such species today.

Once the soil surface is stabilized by drought-resistant mosses and lichens the dunes lose their yellow colour and earn the name of *grey dunes*. At this stage conditions are suitable for colonization by woody plants. At Tentsmuir this includes heathers, both the Common Heather (or Ling, *Calluna vulgaris*) as well as in the damper regions bordering the slacks where the the Cross-leaved Heath (*Erica tetralix*) predominates (see Chapter 5).

Grey Dunes

The natural successor to the yellow dune is the grey dune, where lichen colonisation takes place as the vigour of the Marram Grass declines. Regrettably, these dunes are no longer as prevalent as in times past. Good examples existed at Tentsmuir, but here too they are now declining and being invaded by Rosebay Willow Herb. Extensive damage was caused to the Grey Dunes at Tentsmuir when goats were intoduced in the 1980s in a misguided attempt to reduce an invasion of birch seedlings (see Chapter 10 *Saving the Wilderness*).

Woody plants on sand dunes

The presence of woody plants raises the question of whether or not to encourage the spread of woody species across dune systems. After the disaster at Culloden of the sudden dune migration that took place in 1694 with the extensive movement of the sand dunes, an act was passed by the Scottish Parliament in the reign of King James VII of Scotland in 1695 which specifically *'forbade for hereafter the removal of bent, broom, or juniper, and all woody species.'*

Where grazing from domestic livestock or rabbits is limited, dunes are readily colonised by Pine and Birch, as well as Bent Grass (*Agrostis spp.*) and Broom, with Willow and alder in the wetter intervening slacks. These woody

Figure 4.19 Example of tree growth with mixed species including pine on the landward side of a front dune on the shore near Ventspils in Latvia. Note the spread of roots through the sand from the pine (Photo author).

Sadly, we are too used to considering sand dunes as merely a fringe of grassy hillocks bordering the sea. Less than 0.2 percent of the land surface of the British Isles is occupied by sand dunes (Doody, 2015) and very few of these dunes are completely natural with the progressive development of a succession of dune types including, mobile dunes, fixed and grey dunes, interspersed with slacks and dune heaths as still can be seen at Tentsmuir. In most cases their extent to landward is truncated by roads, car parks or golf courses.

In some areas Pine with Juniper can establish an ecologically attractive forest-cover on dune heaths. Remnants of columnar juniper bushes are still found in a few relict sites in Britain, but it is to the shores of the Baltic that we have to turn to see this vegetation at its best.

When viewed from a European perspective, there are many examples where trees and even coastal forests serve to play a role in reducing sand-dune erosion. Here again, the Baltic has many examples of not only Juniper but also Pine and other woody species colonizing dune systems. Dune systems can have a significant tree presence without the tree cover becoming so dense that it excludes a diverse sand dune ground flora. Figure 4.19 shows pine growing behind the front line of dunes in Latvia on the extraordinary length of the Curonian Spit where they spread their roots widely anchoring the sand while their branches serve to reduce wind speed.

species can increase dune stability by reducing wind-speed at ground level and providing robust root systems which also help to reduce erosion. It will also affect the flora of the understory, as with time the pioneering species of the dunes will be replaced by a more shade-tolerant woodland flora.

The golfers on the Old Course at St Andrews, which lies immediately to the south of Tentsmuir, are well aware of the value of woody vegetation. The recent removal of Gorse and Broom from the links may have facilitated the search for lost balls, but there have been complaints that there is now the irritation of an increase of sand-blow into peoples faces.

The answer as to whether or not woody plants, including trees, have a role in sand dune conservation lies in the contemporary human concept of what constitutes a viable dune system.

Figure 4.20 shows columnar Juniper bushes growing just behind the foreshore dunes on the Estonian coastline. Such a scene may have existed in times past at Tentsmuir in a very similar type of coastal sand dune. A unique photograph of possibly the last single remnant bush of an ancient population of columnar Juniper bushes at Tentsmuir was taken in the mid 1960s where a single mature juniper bush still existed (Figure 4.21). Being a dioecious species this solitary individual (sex unknown) had no chance of reproducing, despite a plea to the Nature Conservancy by the author to find a mate for this solitary plant, nothing was done. Instead, it suffered an ignominious fate as it was removed by contractors working for Scottish Natural Heritage in the belief that it was a merely another Gorse bush!

For many years now Juniper has been absent from Tentsmuir where it could have contributed to the biodiversity of the the duneland flora. Doubtless, there was a reluctance to interfere with Nature. At present Juniper is only found on two dunes in Scotland.

Another woody species that was contributing to the diversity of Tentsmuir was the Sea Buckthorn (*Hippophae*

Figure 4.20 Columnar Juniper (*Juniperus communis*) growing on sand by a Baltic shore in Estonia (Photo author).

Figure 4.21 One last remaining Columnar Juniper bush (*Juniperus communis*) - sex unknown - growing on a rear dune at Tentsmuir as photographed in the mid 1960s. There are now no juniper bushes on Tentsmuir. This specimen was sadly removed as it was mistaken for a Gorse bush! (Photo author).

Tentsmuir's dunes – a changing landscape

Figure 4.22 Sea Buckthorn (*Hippophae rhamnoides*) growing on the dunes at Buddon Ness that lie to the north of Tentsmuir on the opposite side of the River Tay from Tentsmuir (Photo author).

Figure 4.23 Berries of Sea Buckthorn (*Hippophae rhamnoides*). The berries are much sought after by birds. The fruits have long been collected for human medicinal use (Photo author).

rhamnoides). Sea Buckthorn (Figures 4.22-4) as it is called in Britian, is basically not just a coastal plant, as might be suggested from the adjectival part of its common name in English. In German it is called *Sanddorn* which is a more befitting name given its extraordinary geographical distribution on sandy soils from Central Asia through most of Continental Europe.

Sea Buckthorn is common in the more southern and eastern coasts of Britain (Figure 4.24) but has been transplanted to many areas specifically for its ability to resist wind erosion.

When the Gatty Marine Laboratory was first built at St Andrews, Sea Buckthorn was planted along the laboratory walls on the sea facing side for protection, as the neighbouring East Sands at St Andrews are highly prone to erosion. The present managers of this facility have removed the Sea Buckthorn and replaced it with pansies! For a marine biology laboratory this shows a sad lack of ecological wisdom and disregard for the defensive qualities of Sea Buckthorn in exposed maritime situations.

Scottish Natural Heritage has also seen it fit to eradicate this plant from the Reserve at Tentsmuir which has long had well-established colonies of Sea Buckthorn. Whether or not it was intorduced or spread there naturally is not certain (See Crapper 1939). Sea Buckthorn is notable for being one of the few species of flowering plants outside the pea family (Leguminosae) that has the capacity to fix atmospheric nitrogen with the aid of nitrogen-fixing bacteria in its root nodules (Bond, 1955).

Not all Sea Buckthorn bushes have nodules that are active in nitrogen fixation. Tests carried out on the bushes at the Gatty Marine Laboratory by the late

Professor George Bond, (the original discoverer of nitrogen fixation in non-leguminous plants) verified that the bushes at the Gatty Marine Laboratory were active fixers of nitrogen (pers.comm). It is therefore probable that the extensive stand of Sea Buckthorn at Tentsmuir with its vigorous growth was also capable of fixing atmospheric nitrogen which would have been highly advantageous for plants growing on a nutrient-poor sand dune.

Not only does Sea Buckthorn produce dense stands capable of withstanding erosion it also flowers prolifically with the production of much viable seed that is actively spread by birds attracted to the bright orange berries (Figure 4.25).

The decision by Scottish Natural Heritage in 2016-7 to completely remove this species from the Tentsmuir Reserve as it was becoming an impediment to the Fife Coastal Path was perhaps rather an extreme reaction given its usefulness in resisting soil erosion and providing winter feed for a wide variety of birds. Sea Buckthorn is also a useful ecological species as it provides an effective windbreak, and therefore reduces susceptibility to erosion. It also provides a natural input of nitrogen from its nitrogen fixing nodules which can benefit the biodiversity of the ground flora.

Plant survival in sand dunes

The boundary between land and sea is kept in place in sandy shores by plants. Nevertheless, the ability of the vegetation to anchor the coastline has its limits. There is therefore considerable concern that many coastlines may be forced to retreat as a consequence of rising sea levels, increased winter storms, and ever

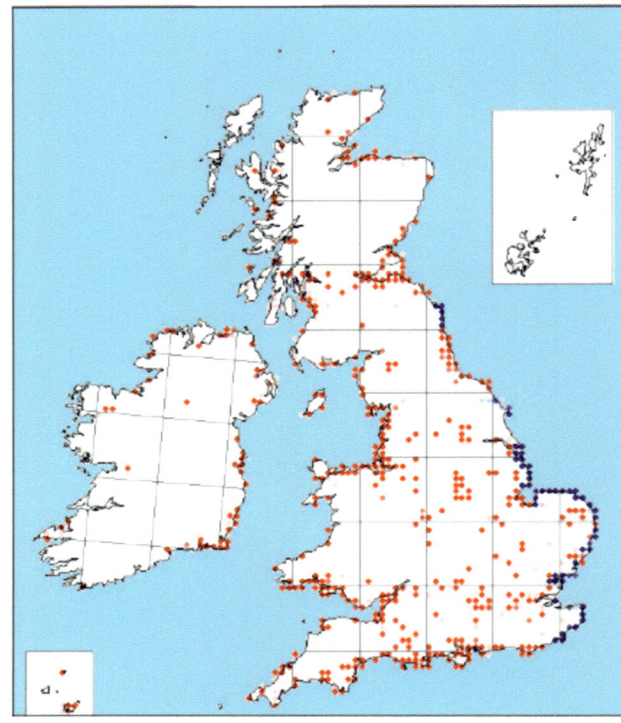

Figure 4.24 British and Irish distribution of Sea Buckthorn. (Reproduced with permission from Preston CD, Pearman A, Dines TD. New Atlas of the British and Irish flora. Oxford: Oxford University Press; 2002.)

more human disturbance. Sea Buckthorn is a species that can be extremely valuable in coastal protection. It is also a species that provides excellent autumn and winter feeding for a wide variety of birds (Figure 4.25). It is therefore difficult either to understand or forgive those who have in recent times, ruthlessly tried to remove this species entirely from the reserve at Tentsmuir.

Drought tolerance

Tolerance of drought is an outstanding feature of sand-dune vegetation. Many of the moss and lichen species which secure the surface of the sand dunes have a

Figure 4.25 Birds enjoying feeding on Sea Buckthorn berries. (a) Robin, (b) Blue Tit, (c) Mistle Thrush (d) Fieldfares (Photos Professor Stephen Buckland).

remarkable property of being able to allow their tissues to dry out without losing viability. After prolonged periods of desiccation (e.g. lying on a herbarium sheet for 70 years!) some moss species begin to resume metabolic activity within 30 minutes of gaining access to water. One of the dangers of desiccation injury, especially when plants are exposed to sunlight, is the generation of highly destructive oxygen free-radicals by transfer of energy from excited chlorophyll to oxygen. The dune-moss *Tortula ruraliformis* when desiccated in the light contains high concentrations of the defensive anti-oxidants α-tocopherol and glutathione which contribute to its remarkable desiccation tolerance (Seel *et al.*, 1992).

Little thought is given to the needs of sand dune flowering plants for water, probably because most perennial dune flowering plants appear to be either economical with their water supplies or else restrict growth to seasons when water stress is not a serious problem. Dunes have a characteristic flora of winter annuals with species such as Common Whitlow Grass (*Erophila verna;* a diminutive member of the cabbage family) and Spring Vetch (*Vicia lathyroides*) which survive the heat and drought of summer as seeds, then germinate in the autumn, grow over winter and flower and seed in spring and early summer the following year.

Marram grass (*Ammophila arenaria*) the characteristic species of dune-dune tops with its hard in-rolled leaves appears to be the epitome of drought-resistance. Not only is transpiration reduced to a minimum, but the deep root systems are able to access water from the lower moist layers in the dunes.

Water is always an essential resource, and is regularly in limited supply even for the needs of Marram in dunes. The transpiration needs of dune vegetation can exhaust the rain water held in the rooting zone in a typical sand dune in four days (Salisbury, 1952). Nevertheless, it is obvious that dune plants do not wilt after 4 days drought, which raises the question as to how they obtain their water. The elevation of water by capillary action will not move water more than 40 cms above the water table. As the average rooting depth of Marram Grass is in the order of 1-2 metres and the depth of the water table can be 6 metres or more below the surface of a high dune, capillary action alone is an inadequate explanation for the replenishment of water supplies in large dunes.

A significant upward movement of water does take place however by internal condensation. At night, as the upper layers of the sand cool, the lowers layers remain warm. Consequently, water vapour moves upwards and condenses in the cool upper layers of the sand – a process described as *internal dew formation*. It has now been found that the deeper-rooted plants also contribute to the water supply of the upper layers of the soil by a phenomenon described as *hydraulic lift*. During the night the deeper-rooted species release some of the water they are transporting upwards into the top layers of the soil where it becomes available to shallow-rooting species (Williams *et al.*, 1993).

The shallow-rooted species such as Birds Foot Trefoil (*Lotus corniculatus*), Clover (*Trifolium spp*) and Rest Harrow (*Ononis repens*) profit from the presence of deep-rooted plants by the provision of indirect access to the ground water reserves that would otherwise be beyond their reach. The deep-rooted plants in turn benefit from the growth of the shallower rooted plants as many of them can fix atmospheric nitrogen (Bird's Foot Trefoil, Clover, Medick) and thus contribute to the nutritional status of the entire plant community.

Conclusions

The ability of the vegetation to anchor the coastline and respond to physical and environmental change has its limits. There is therefore considerable concern that many coastlines may be forced to retreat as a consequence of rising sea levels, increased winter storms and human disturbance. Given that we are powerless to reduce the tendency of the sea-level to rise, or abate any increase in storms, protection measures for our shorelines need to have a biological basis. Physical shore defences, whatever they are made of, be it wood or concrete, only serve to sterilize the shoreline. For long-term protection of the coast there has to an understanding of the natural stresses that threaten plant survival in fragile coastal habitats such as Tentsmuir.

Sand dune systems in their natural condition with their varied lines of defence, fore dunes, yellow dunes, grey dunes and slacks, are collectively well-adapted to withstand disturbance from the sea. Most sandy beaches can repair storm damage, provided the detritus of the sea, seaweed and other flotsam, are not removed and there is an adequate supply of sand and the disturbance is the habitual frontal attack. Beaches are nevertheless vulnerable to an attack from the rear and unexpected alterations in their environment. Although sand dune vegetation is highly drought resistant, the water table must be accessible.

Falling water tables, removal of expected resources such as seaweed, truncation of their natural development by roads and golf courses, all contribute to weakening the resilience of dune systems to withstand physical disturbance, whether from human interference or natural disasters. The conservation of biodiversity in dune systems however, requires more than just ensuring the physical preservation of the dune.

Over-protection can also result in habitat diversity being reduced and result in the habitat becoming homogenous with the result that species diversity is lost. For long-term preservation of dune systems and their biota a careful balance has to be struck between ensuring adequate physical protection yet still enabling the cycles of denudation and recolonisation to take place so that the full range of natural habitats and their species are preserved.

Chapter Five

Tentsmuir's wetlands

Dune slacks and salt marshes

One of the most powerful discriminating factors affecting plant distribution is tolerance of flooding. The flora of wetland habitats differs distinctly from that which is found on land elevated above the floodline and free from the risks of inundation. This distinction is noticeable even if the duration of the flooding is only for a few weeks in the year. The longer the period of inundation, the more specialised becomes the flora of the flooded area.

Tolerance of flooding clearly demands certain special adaptations to enable plants to withstand the exclusion of air from the soil. Such a situation can be overcome in two ways, either by evolving a metabolism that can function without oxygen for a limited period (e.g. winter), or else to have a means for supplying air to inundated roots.

The vegetation of Tentsmuir's wetlands comprises two further distinct types in relation to flooding depending on whether or not the inundation is from salt or fresh

Figure 5.1 The Sea Club Rush (*Bolboschoenus maritimus*) (a) growing in brackish water (b) a rhizome of this species that has been kept in an anaerobic incubator without any access to oxygen for 12 days during which time it has grown a healthy new shoot which then survived being restored to normal air (Photo author).

water. The most specialized plants are those species that can endure both the exclusion of oxygen and exposure to salt as a result of flooding by sea water.

Resistance to flooding has earned many plants specific latin names such as *aquaticus, fluviatilis, inundatutus, palustris etc*. Equally descriptive are whole groups of wetland plants being described in vernacular terms such as rushes, reeds and sedges. Tolerance of flooding clearly demands that flowering plants require special adaptational strategies. Among the many species of the wetland rushes it is the Sea Club Rush (*Bolboenus maritimus*), which is found on the north shore of Tentsmuir (Figure 5.1), is especially remarkable in that it can not only survive flooding but can also grow new shoots even while totally deprived of oxygen. Such is the adaptability of this species that it can survive a prolonged and total absence of oxygen (anoxia) provided the oxygen supply is eventually restored. Many other wetland species can endure a partial reduction in their oxygen supply (hypoxia), which can also be for a prolonged period.

Indirect harmful effects of inundation can also arise from toxic substances that accumulate in flooded soils such as ferrous iron, sulphides and manganous ions. Such damage, even if it only causes minor injuries can predispose plants to fungal infections and attacks by nematodes and other parasites.

Despite the hazards to plant life that arise in wetlands, evolution has created a variety wetland floras that have been able to adapt to flooding. The differing ways that this has been achieved has created an enormous diversity in the plants of bogs, and marshes. Even within the confines of Tentsmuir, there exists a wide range of wetland habitats which show considerable diversity in their flora due to the different patterns of flooding in terms of frequency and duration and whether inundation is by salt or fresh water.

In dunelands these flood-prone areas are commonly referred to as *slacks* The word *slack* (cf. *slake*, to allay thirst, to make wet e.g. *slaked lime*) implies a tendency for these areas to be flooded particularly when the water table rises in winter and spring.

The basis for this diversity can be studied by imposing plants experimentally to varying types of flooding one by one. To investigate the consequences of an interruption to the oxygen supply plants can be put in anaerobic incubators from which oxygen is totally excluded. In this somewhat extreme environment most flood-sensitive species die within a few days.

In wetland species considerable variation can be found in the length of time that plants can be deprived of oxygen without dying. Species such as the Yellow Flag Iris (*Iris pseudcacorus*), and the Sea Club Rush (*Bolboenus maritimus*) both of which can be found at Tentsmuir and can endure lengthy periods of total deprivation of oxygen, often in a dormant condition and then resume growth on restoration of access to air. Some species however die on restoration of access to oxygen, a phenomenon that is similar to the post-cardiac arrest injury in human beings.

Avoidance of anoxic stress

A wetland species commonly found at Tentsmuir is Reed-Sweet Grass (*Glyceria maxima*). This species is remarkable in that it combines a limited tolerance of

species of wetland habitats and found at Tentsmuir. The common name – Meadowsweet does not come from the term for a hay meadow. Instead, it reflects the ancient use of this plant to sweeten *mead* - cf. German *Mädesüss*).

Physiologically, experiments show that Meadow Sweet although being more tolerant of oxygen deprivation than Reed-Sweet Grass (*Glyceria maxima*) can however be displaced from the wettest habitats by the less

Figure 5.2 Reed-Sweet Grass (*Glyceria maxima*) growing at the rear of the slack on the seaward side of the Ice House see Figure 5.3)

anoxia with a capacity to supply its roots with oxygen even when flooded (Figure 5.2). This is achieved by having basal shoots that remain green in winter and consequently are photosynthetically active even when under water. As the inundated leaves can carry out photosynthesis when flooded they generate an internal source of oxygen which alleviates the lack of soil aeration. This over-winter photosynthetic capability also enables this species to make an early resumption of growth in spring before the winter flood-waters have fully subsided which gives this species significant advantages in relation to competition with those wetland species that have to wait until the flood waters subside before they can resume growth.

Queen of the Meadow (*Filipendula ulmaria*) also commonly called *Meadowsweet*, is another common

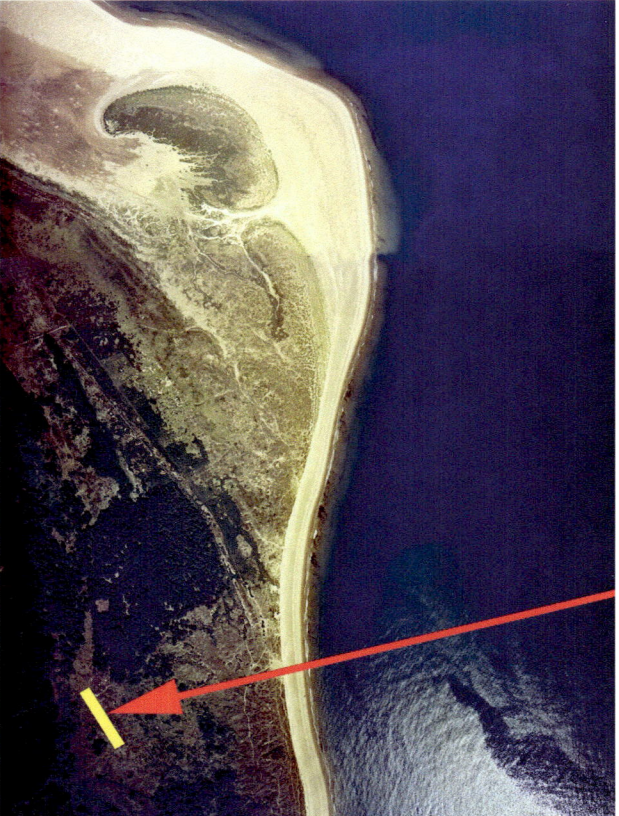

Figure 5.3 Aerial view of the northern region of Tentsmuir showing the position (yellow band) of the mapped rear slack that was monitored in relation to flooding frequency over 24 years from 1964 to 1986

frequently (Figure 5.3). This transect is still in place and is regularly revisited (see Figures 5.4-6).

In the late 1960s and early 1970s a new ditch from the forest was created by the Forestry Commission in an area just to landward of the rear slack (Figure 5.3).

As a result the drainage of this ditch during periods of flooding, more water decanted onto the rear slacks which flooded with greater frequency and for longer periods in winter (Figures 5.4-5).

The detail of the relative positions of Reed Sweet grass (*Glyceria maxima*) and Meadowsweet (*Filipendula ulmaria*), before and after the increase in flooding can be seen when comparing their location in Figures 5.5 a-b. It can be seen in these maps of 1964-88 that the low ground has been abandoned by the Meadowsweet and replaced by its more successful neighbour Reed Sweet grass

These changes in the respective distribution of these two species reflect their differing responses to flooding as observed when subjected experimentally to controlled flooding and oxygen deprivation. Expressed in human terms the over-cautious yet highly flood-tolerant Meadowsweet lost ground, not from being intolerant of flooding, but because it did not resume spring growth until the flood-waters had subsided. Consequently, this early empty space in spring was taken over by the more precocious Sweet Reed Grass.

The most stable boundary on the transect was at the southern end where at the boundary between between the Sand Sedge (*Carex arenaria*) and the

Figure 5.4 Winter view of flooded rear slack at Tentsmuir photographed in winter 1975. Note posts in distance marking the line of the transect.

anoxia-tolerant Reed Sweet Grass due to an inability to resume growth before the flood waters have totally subsided. Due to this tardiness in resuming growth in spring Meadowsweet can be out-competed by the less anoxia-tolerant Reed-Sweet Grass. This is clearly seen in some of the wettest parts of the dune-slacks at Tentsmuir (Figures 5.3-4).

In 1964 a transect designed to study the relative flooding tolerance in some of the varying plant communities in Tentsmuir was set up in a wet-dry transition area to the east of the Ice House (Figure 5.3). The transect ran from a relatively elevated area with non-flooded dune heath northwards to the edge of the Powie Burn which was an area of wetland vegetation that flooded

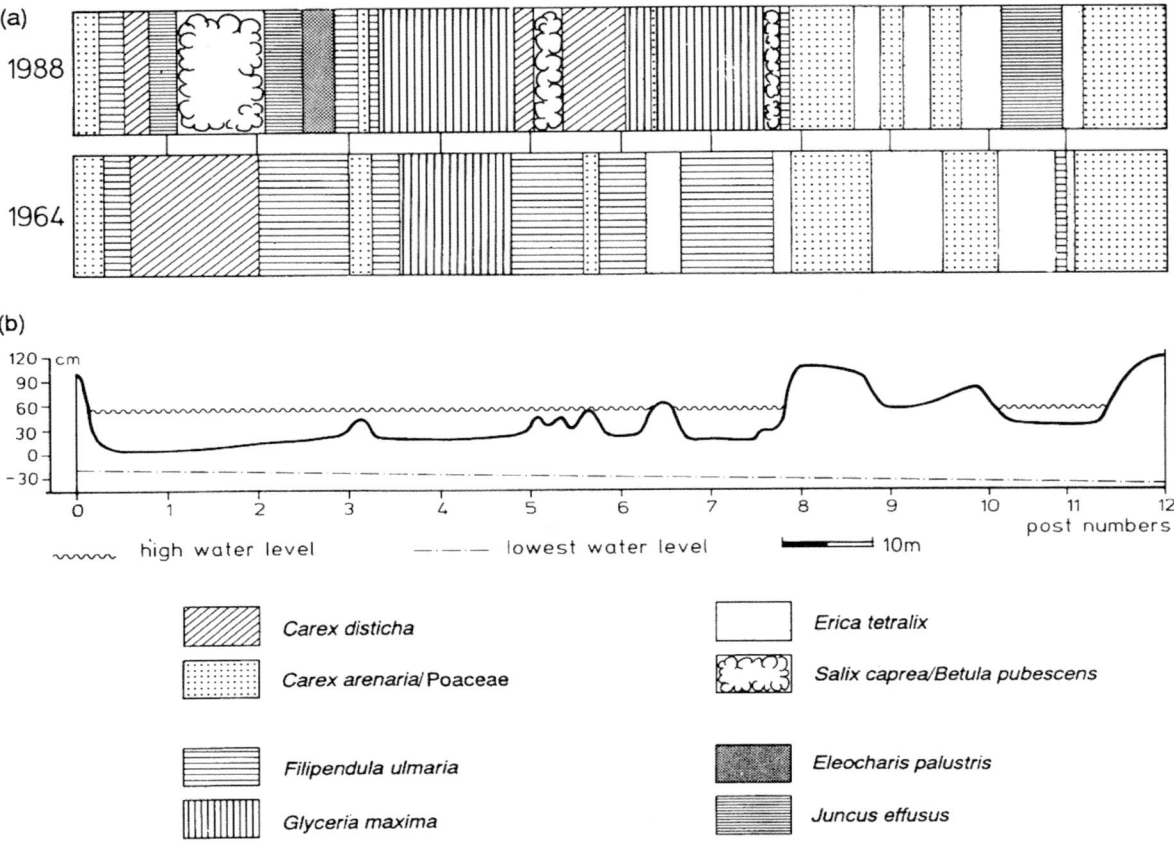

Figure 5.5 Maps of a permanent dune slack transect in an area prone to prolonged flooding (see Figure 5.4) Transect mapped (a) initially in 1964 and then repeated (b) after a 24-year interval in 1988. Flooding in this area is now in 2018 much reduced. Map reproduced with permission from Studer-Ehrensberger et. al. 1993).

heath vegetation of the dune tops lies with the Meadowsweet dominated community below. This boundary coincided with the maximum winter floodline which has remained stable during this entire period, as at this point the excess water decants onto drier slacks to seaward (Figure 5.5).

The dune and slack area at Tentsmuir has no permanent open stretches of water. However in wet years, one of the more tolerant semi-aquatic species Water Crowfoot (*Ranunculus aquatilis*) can be found where the flood-waters lie longest. Most typically, this is in water-logged ditches and other low lying areas near the water table in the wetter parts of the wet slacks. This species can survive when not flooded but is usually only obvious when it flowers in wet years especially when there is prolonged spring flooding (Figure 5.6.).

Dune-slack variation

The variety of responses of the plant species to flooding at Tentsmuir, whether it be by salt or fresh water create considerable biodiversity in the vegetation as it develops to suit particular locations. Within the boundaries of Tentsmuir the wetlands comprise regions with both salt and fresh water flooding. The areas exposed to flooding with sea water are usually described as *Salt Marshes* and the fresh water areas as *Slacks*.

Flood-line alder association

A long-established and conspicuous feature on the Tentsmuir dune system, was the *flood-line alder association*. This formation arises when alder seeds are washed up on the edges of the slacks (Figures 5.7-8).

From dendrological trunk core borings made in 1966 the age of these alder trees at that time was estimated to lie between 20-25 years. The alders in the southern part of the reserve have now been largely removed or killed by salt flooding. By contrast, those in the northern region that still remain are now about 75 years in age and have not as yet suffered such damage.

Figure 5.6 Water Crowfoot (*Ranunculus aquatilis*) growing on late flooded ground at Tentsmuir.

Figure 5.7 Remnants of the once extensive food-line Alders to the north in Tentsmuir.

Figure 5.8 Remains of the more northern Alder trees as a result of erosion and salt damage

The terrain to either side of the Alder trees was in the past one of the richest at Tentsmuir in terms of species diversity. The floristic richness found there in the past included many orchid species but is now greatly diminished, partly due to erosion and probably also due to a general drying out of the slacks. (See also Chapter 6).

Salt slacks

The Sandwort (*Honkenya peploides*) community that now dominates the salt slack at Tentsmuir Point differs from that which was there 50 years ago. In the 1960s the salt slack lay 2-3 ft (60-90 cm) above the high water mark. Erosion has now left a salt slack that is much closer to the high tide level and consequently there is much bare sand (Figure 5.9.) The pale yellow patch in the top left hand portion of Figure 5.9 is part of this present salt slack.

In this location, the slack vegetation is continually prone to flooding with sea water as well as erosion. In the past, although this plant community was similar to that of today, the dominant role of Sandwort was shared with other halophytic species The former extent of this slack in 1965 covered an extensive area which was located seawards of the anti-tank blocks which were laid down in 1940. Due to the massive erosion that has taken place recently at Tentsmuir Point the tank blocks have now largely disappeared along with much of the salt slack vegetation (Figure 5.9).

At present the salt marsh is still susceptible to erosion and although it covers an extensive area it is also highly exposed to salt-flooding. Nevertheless, it does attract the cattle that graze on the nature reserve in summer (Figure 5.12). The flora of salt tolerant plants (*halophytes*) now present on the slack is not as rich in species as it was 50 years ago. Nevertheless, Sea Milkwort (*Glaux maritima,* Figure 5.13) is still found in this region although less common than in the past.

Figure 5.9 The *Honkenya peploides* (Sandwort) slack as it is now (2018). Erosion has replaced the former slack of fifty years ago when this area was 2-3 ft. (60-90 cm) above the high water mark. The present Sandwort community is the pale yellow patch in the top left hand portion of the picture. See inset for detail of these plants.
Inset: A vigorous growth of Sandwort, detail of flower.

Figure 5.10 Dense patches of the present salt marsh vegetation consisting mainly of Sandwort (*Honkenya peploides*) which in places has developed sufficiently to provide attractive grazing for cattle (see Figure 5.12).

Figure 5.11 A vigorous growth of Sandwort with inset showing detail of flower

Figure 5.12 Cattle grazing in late summer on a salt marsh. At this time of the year the salt-tolerant plants have developed high concentrations of sugars as a defence against excessive salt uptake and therefore make attractive grazing

Lotus corniculatus slack

The next type of slack to develop in the sequence beginning from the shoreline characterised by the presence of Birds Foot Trefoil (*Lotus corniculatus*) and Creeping Willow (*Salix repens*). Both these species are still dominant and are highly visible across the Great Slack. This now largely desalinated former salt slack type which in 1965 lay just to the former landward side of the line of coastal defence concrete blocks which are now no longer present in this area as it has suffered enormous erosion. These

Figure 5.13 Sea Milkwort (*Glaux maritima*) growing in the coastal slack sometimes referred to as the Great Slack at Tentsmuir.

Soils in this area have typically a high pH and are rich in sodium and potassium. The chloride ion content becomes reduced with distance from the sea as the sandy soil does not retain anions and they are readily leached out by summer rains. As the slack is only marginally above the high water mark, the drainage seawards is slow. Nevertheless, the water beneath the top soil contains considerable quantities of chloride.

In 1965 the area was described as a salt slack and lay sufficiently above the high tide level to contain a number of species that are not generally classified as salt-tolerant (halophytes). A notable example of this was Yellow Rattle (*Rhinanthus minor* – Figure 5.12*)*. Yellow Rattle is a semi-parasite, which mainly attacks grasses and is still found in the salt slacks along with Red Fescue (*Festuca rubra*) which it frequently parasitizes.

Figure 5.14 Yellow Rattle (*Rhinanthus minor*) a common parasite of grasses.

slacks can be distinguished from *Honkenya peploides* slacks by being found on ground that is normally normally just above the high water mark.

In the survey of the dune system carried out in 1965, the slacks dominated by Birds Foot Trefoil (*Lotus corniculatus*) were found to lie about 8-10 ft (250-300 cm) above the high tide level. This extra height gave a better drainage gradient towards the sea and were therefore free of the large concentrations of chloride found in the soil where Sea Sandwort (*Honkenya peploides*) dominated. Typical also of this areas found at that time was the Northern Marsh Orchid (*Dactylorhiza purpurella* Figure 5.15).

To the rear of these slacks lay the region where the Flood-Line Alder Association developed as a striking feature of the vegetation half a century ago (Figures 5.7-8).

Erica tetralix slacks

Further inland on slightly higher ground, on the eroded remains of ancient dunes and just seawards of the marsh zone and the region of maximum flooding, a series slacks can still be found where the Cross-Leaved Heath (*Erica tetralix*) is a dominant feature (Figure 5.16). Three types of slacks are characterized by the presence of Cross-leaved Heath (*Erica tetralix*). In the wettest ares it occurs along with Meadowsweet (*Filipendula ulmaria*). In areas a little less prone to flooding it is found with Red Fescue (*Festuca rubra*). A third combination with the lichen *Cladonia sylvatica* has created an *Erica tetralix* slack which is less exposed to inundation.

Figure 5.15 The Northern Marsh Orchid (*Dactylorhiza purpurella*) growing on the edge of the Great Slack in the mid-1960s.

Despite the fact that all these communities are found on a uniform sandy soil, it is quite remarkable that so many different patterns of vegetation have developed due entirely to differences in salination, and inundation. Equally remarkable, is the number of species that have colonized these different habitats and created such an ecologically diverse range of wetland vegetation types.

The desiccation of the slacks

Viewing the changes that have taken place in the flora at Tentsmuir over the past 50 years it is evident that it is the dune-slacks that have suffered the greatest impoverishment in terms of plant species. Over this period the frequency of winter-flooding of the slacks has noticeably decreased. It would therefore appear

Tentsmuir's wetlands 81

Figure 5.16 Two of the three types of *Erica tetralix* (Cross Leaved Heath) slacks seen here by banks of the Powie Burn. In the foreground on land prone to flooding is the type associated with *Festuca rubra*. In the middle distance on higher ground the pale colour of the lichen *Cladonia sylvatica* can be seen with *Erica tetralix* on land that is less likely to be flooded.

that it is the desiccation of the slacks that has mainly brought about this loss.

Seventy five years ago a local naturalist living at Tayport, Ellis Crapper, who had been studying the flora of Tentsmuir since the early 1900s predicted that this species richness which he described as '*the glory of Tentsmuir* would be likely to diminish as the new forestry plantation planted there after the First World War matured. It is now clearly evident that his pessimistic foresight was correct.

The scale of the change that has taken place in the environment at the north east part of Tentsmuir can best be gauged when viewed from the air (Figure 5.18). The nature reserve in particular in its coastal fringe locality is entirely dependent on the water that flows through the sand that lies below this extensive forestry plantation. A forest of this size and maturity will remove a very considerable proportion of the water that is available in the area through evapotranspiration from its foliage. To this must be added the extended drainage that has been put in place to facilitate the root penetration of the forest trees in an attempt to reduce the ever present risks of wind-throw, when deeper roots fail to develop due to a high water table.

The vegetation in the fresh-water slacks is highly dependent on having access to large reserves of fresh water below the soil surface. An early study of the effects of tree planting on the level of the water table was able to detect significant reductions within 11 years of planting (Ovington, 1951). It is now obvious that it is the dune slacks with their unique flora combining species from both more northern and more southern regions that is being ecologically impoverished and their ancient floristic heritage is being lost as a result of the proximity of this very substantial forestry plantation (Figures 18-19).

A comparison of the vegetation of the rear slack just to the east of the Ice House described above as it was in 1975 with what is there now in 2018 shows the extent of this change (Figure 5.17).

Figure 5.17 Example of reduction in flooding. The long-term monitored rear slack as it is at present (2015) See inset of same site in 1975 (and Figure 5.4).

Figure 5.18 Aerial view showing the extensive area of Tentsmuir Sands that is occupied by forest as compared with the coastal fringe occupied by the nature reserve. The evapotranspiration of water from the forest reduces the level of the water table and deprives the nature reserve of the water that is particularly important for the maintenance of biodiversity in the dune slacks (Photo SNH).

It is now abundantly clear, as was forecast by Ellis Crapper (Crapper, 1940) that the floristic richness of the Tentsmuir flora has become impoverished by the maturation of the Forestry Commission plantation (Figure 5.19).

Some remedial actions are clearly needed and these are discussed in Chapter 6. This will include an argument for the re-introduction of the Beaver in an attempt to raise the level of the water table. There is currently active discussion of the need for the *rewilding* of some of our cherished Scottish landscapes. Such schemes do not need to be solely based on dramatic reintroductions such as the Wolf, or the Lynx. A more skilful management of water tables, could with planning and imagination, create an environment that would provide an income for forestry as well an ecologically enriched environment.

The Tentsmuir forest is a suitable area to develop this argument. Tree productivity is limited in this area due to the combination of sandy soils and water table fluctuations. The drainage is barely sufficient to avoid the risks of wind-throw in this coastal habitat. Consequently, a considerable part of the forest income comes from the car park charges for the visiting public. If the public could be treated to an ecologically enriched forest, with the possibility of a view of one or two beaver lodges it could easily provide in an area adjacent to a major city (Dundee) an environmental attraction as well being viable forestry plantation.

Figure 5.19 A Beaver lodge in a Russian wetland forest – a possible future visitor attraction at Tentsmuir which might also help to restore the ancient wetland vegetation (Photo F. Fyodorov).

Chapter Six
Land, people and resources

Tentsmuir as a sporting estate

In Medieval times, Tentsmuir served as a Royal Hunting Forest (see Chapter 2) where hawking would have been pursued and property rights clearly defined in the various royal charters that were granted from the eleventh century onwards. Tentsmuir has therefore had a long history of human involvement and interest, both in terms of its ownership, sporting rights and wild life.

Tentsmuir ornithology

Regular documentation of bird life at Tentsmuir is due largely to the Berry family who have lived continuously at Tayfield House for 7 generations since their ancestor, John Berry bought the greater part of the Inverdovat Estate overlooking the Tay Estuary in 1787. Mr William Berry (4th of Tayfield, 1864 -1954, Figure 6.1) had an especial interest in the birds of Tentsmuir and brought many kindred spirits to the area in the closing years of the 19th and the early years of the 20th century.

Probably the most notable among them were Dr. Harvie-Brown (Figure 6.2) and Dr. William Eagle Clarke (Figure 6.3). Both provided accounts of birds in Scotland which included details on Tentsmuir. They both also received honorary degrees for their studies of the flora and fauna of Scotland, Dr. Harvie-Brown (LL.D University of Aberdeen 1912) and Dr. Eagle Clarke (LL.D University of St Andrews 1916).

Figure 6.1 William Berry 4th of Tayfield (1864-1954) whose interests in bird populations and conservation management brought many ornithologists to Tentsmuir. (Photo courtesy of William Berry 6th of Tayfield).

Figure 6.2 Dr. J.A. Harvie-Brown (1844-1916) distinguished ornithologist who took an active part in developing the study of bird migration in the latter years of the 19th century. (Photograph from his 1905 book book *Travels of a naturalist in Northern Europe*).

travelled widely in Norway and Russia. His most adventuresome journey in 1875, to the lower reaches of the Pechora river together with Henry Seebohm (author of *The Birds of Siberia*) is recorded in his own book on his Russian travels. *Travels of a naturalist in Northern Europe*, (Harvie-Brown, 1875). Harvie-Brown had a very extensive and accurate knowledge of birds and their habits, and was particularly interested in problems of distribution and migration. In relation to Tentsmuir his book, *A Fauna of the Tay Basin and Strathmore* (1906), is a mine of local information.

Dr. Eagle Clarke (Figure 6.3) was a Yorkshireman from Leeds, where his father, William Clarke was a solicitor. He studied at Yorkshire College, Leeds and initially worked as a civil engineer and surveyor, but later took up natural history as a profession. Dr. Eagle Clarke became Curator of Leeds Museum in 1884, moving to the Natural History Department of the Royal Scottish Museum in 1888, where he was Keeper from 1906 to 1921. He had a life-long interest in bird migration, and was instrumental in recognising that lighthouses and lightships were capable of recording scientifically important information on bird migration.

The great wealth of bird-life at the mouth of the Eden, and along the sandy shores of Tentsmuir to the river Tay, was described by J.A. Harvie-Brown in his classic work *A Fauna of the Tay Basin and Strathmore*. This particular area he recorded as a, 'cheerful sight for the wandering ornithologist.' Dr. Eagle Clarke, also noted the richness of bird-life at Tentsmuir, 'notwithstanding the persecution they receive from the punt shooters and shore hoppers of Dundee and St Andrews.' He noted that that 'the land was severely raided in the

Figure 6.3. Dr. W. Eagle Clarke (1853-1938), pioneer of studies on bird migration, and a frequent visitor to Tentsmuir.

Like Charles Darwin in his youth, Dr. Harvie-Brown spent much time shooting game birds, and like Darwin, this led to a more general interest in birds. Harvie-Brown was born in 1844 at Dunipace where he later inherited a 2100 acre estate. He studied at the Universities of Edinburgh and Cambridge and

food supply appropriate measures were taken. This care for the food supply also involved planting heather and sowing seeds of a suitable range of species that would produce feeding for the birds that came there in winter, such as the Snow Buntings, which still continue to be one of the winter delights of Tentsmuir (Figure 6.4.)

In addition to protecting the wildfowl populations at Tenstsmuir William Berry also created a grouse moor (see below). The protection given to the Grouse, proved to be beneficial to other wild birds and notably increased the number of nesting Shelduck and Eider.

Bird protection, in this part of Tentsmuir, also included the control of Crows. However, Owls, Kestrels and Merlins and the larger raptors that visited from time to time were carefully protected.

Due to these early conservationists, subsequent generations of ornithologists have long been able to appreciate the uniqueness of the bird populations at Tentsmuir. This is well-recorded in the pioneering work, for long the standard book of reference for Scottish birds, written by the distinguished lady ornithologists Evelyn Baxter and Leonora Rintoul (Bauch and Berndt, 1973).

Both these ladies were well acquainted with Tentsmuir. Figure 6.5 records their visit there in 1912 on the occasion of a picnic held at Morton Lochs to celebrate the 5th birthday of. John Berry, who later in life was to become the first Director the Nature Conservation Council in Scotland and create Tentsmuir as its first Reserve in 1954.

Figure 6.4 Snow Bunting. One of the species of birds that were already being encouraged at Tentsmuir during the shooting tenancy of Mr. William Berry in the early 1900s and have still a regular and outstanding presence at Tentsmuir in Winter. (Image Len Fullerton with permission of the artist's daughters)

nesting season by bands of boys that come across from the other side of the Tay and take all the eggs they could find upon the Tents Muir and shoreline.' He recorded, however, that 'the passing of the Bird Protection Acts of 1894, which he noted, '*were carefully drawn up for specific areas for Fife County Council by men such as Mr. William Berry, who were in the best position to judge local needs, were proving highly effective.*'

William Berry (4th of Tayfield) operated a private bird reserve on the northern part of Tentsmuir on Shanwell Muir where he was the shooting tenant from 1901 up to the outbreak of World War I. Great care was taken to create habitats that encouraged as large a variety of birds as possible. The Shanwell shooting was initially a duck-shoot. This was, however, done with due regard for the balance of the species present. When Mallard and Teal became too numerous for the

Figure 6.5 Picnic at Morton Lochs in August 1912 to mark the 5th birthday of John Berry the future first Director of the Nature Conservancy Council for Scotland and the eventual creator of its first National Nature Reserve at Tentsmuir in 1954. Young John is sitting on the right holding in his hand what is possibly a young owl. Facing the camera is Leonora Rintoul. Her future co-author of The Birds of Scotland (1953), Evelyn Baxter, is seated behind and to her left (Photo by courtesy of William Berry 6th of Tayfield).

Tentsmuir in the late 19th century was shared by three estates, Scotscraig and Shanwell in the north, Kinshaldy in the central region and Earlshall in the south (see below Figure 6.7). Stray Grouse may have crossed the Tay from time to time, but never established themselves on Tentsmuir as a breeding population. One bird, a hen, was killed in 1872 by John Berry (3rd of Tayfield). The Gamekeeper on the Scotscraig Estate, John Fowlis (Figure 6.6) was also aware of other occasional birds (Berry, 1894).

The Tentsmuir Grouse Moor

At the time of this early 20th century picnic (Figure 6.5) Tentsmuir was, as it had been for many years, an open moorland frequented by coastal birds, as well as being a traditional duck shoot. For a short period from the 1890s to the outbreak of the First World War it was also managed as a grouse moor.

An account of the introduction of Grouse to Tentsmuir was given by Dr. John Berry's father William Berry (4th of Tayfield). Although this moorland at its northern extremity appeared at that time to be suitable for grouse, yet no grouse had become established there naturally (Berry, 1894).

Figure 6.6 John Fowlis who was noted as the gamekeeper who successfully managed the introduction of grouse to Tentsmuir (Photograph from Harvie-Brown 1906).

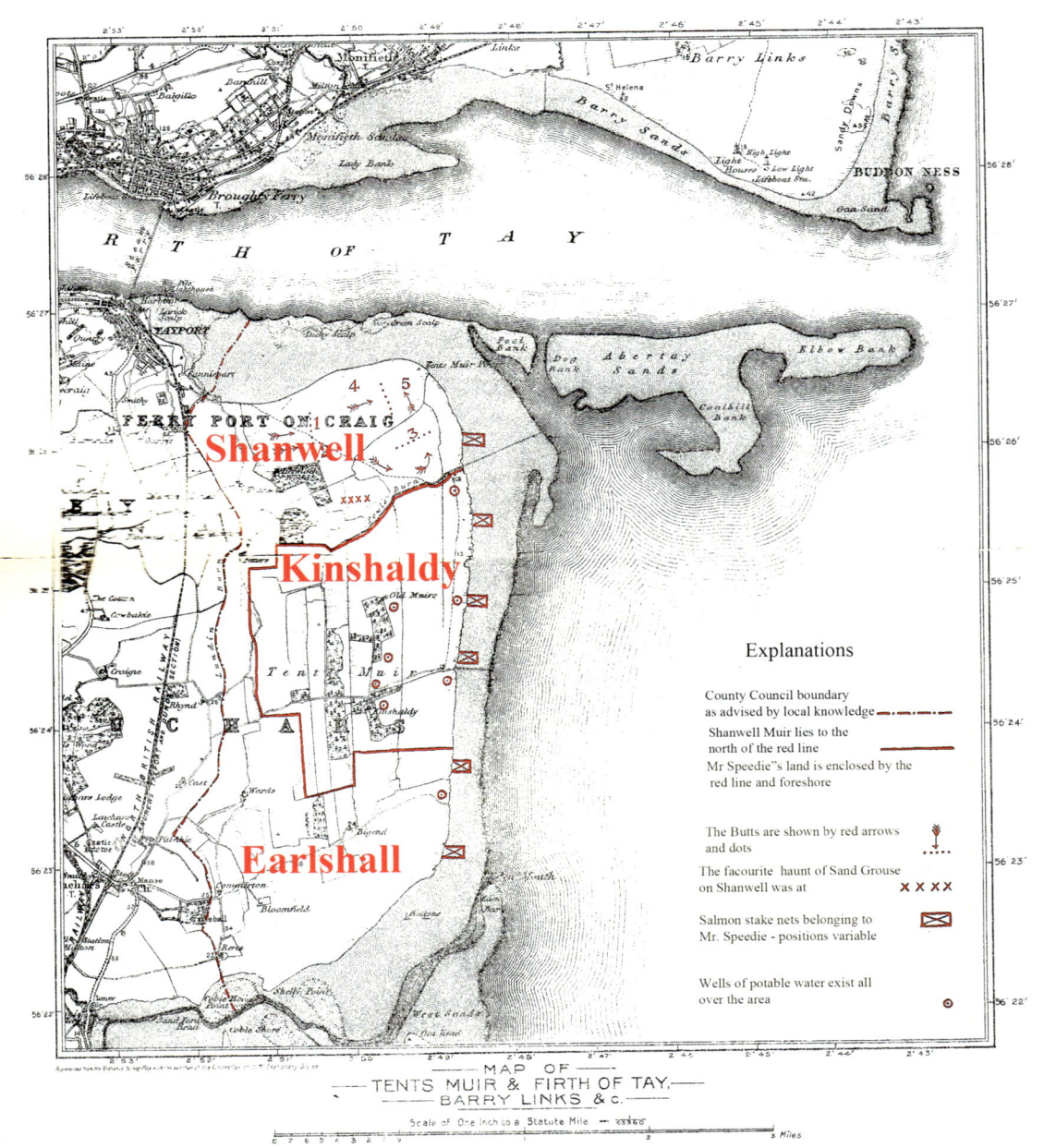

Figure 6.7 Map of Tentsmuir, the Firth of Tay and Barry Links as prepared for Harvie-Brown (1906) by William Berry 4th of Tayfield with clarification of the explanations and labelling of the major divisions of land on Tentsmuir as they existed when the map was prepared in 1906.

Land, people and resources

The northern part, Shanwell was part of the Scotscraig estate of which William Berry 4th of Tayfield was the shooting tenant. It had not escaped the notice of the gamekeeper, John Fowlis, that Grouse regularly fed on the Creeping Willow (*Salix repens*) which was then, widespread in the northern part of Tentsmuir until largely removed by coastal erosion in recent years (see Chapter 3).

Much of the northern area, was also heather-clad at this time, with intervening dry hollows which distinguished it from the more southern parts which were more typically wet between the dunes and suffered frequent flooding. These so-called *dune-slacks* often remained flooded to a depth of 6-8 inches from late autumn until spring. This northern region of Tentsmuir, although somewhat different from traditional grouse moors, had nevertheless, many aspects which suggested it might prove suitable for grouse.

In 1876 Admiral Maitland Dougall from neighbouring Scotscraig, decided to try an experimental introduction of some Grouse. On 2nd of July, John Fowlis, the Scotscraig gamekeeper, with the keeper from Blairadam went to Outh Muir (Fife) and transported two well grown young Grouse to Tentsmuir where they were set at liberty.

This same process was repeated on 8th of August the following year when 8 Birds were introduced to the 1,000 acre portion at Tentsmuir Point. Particular credit for the success of this introduction of Grouse to this north-east corner of Fife was given to John Fowlis, the long-term gamekeeper to the shooting tenant there at that time (Figure 6.6).

It appeared that the experiment was proving successful. Great care was taken in this dry sandy environment to provide the birds with adequate water. Water holes were therefore dug to facilitate the survival of the birds in this arid region of the peninsula. In addition, a large staff of men was sent down to dig up turves and then plant them in suitable spots where Bell Heather (*Erica cinerea*) or grass predominated. This appeared to be a success for on 26th of August 1878 five birds from this first introduction were shot by Admiral Dougall (Berry, 1894).

Great care had to be taken against marauding crows. However, the grouse population rose steadily and it was thought therefore worthwhile to erect driving butts. By 1890 several butts had been erected (see Figure 6.10) and on one occasion, a total of over 73 brace was reached. The moor was never claimed to be the most prolific of grouse moors, possibly due to its coastal location.

Figure 6.8 Grouse drive in progress at the north end of Tentsmuir on the 21st September 1904 with cadets from the Mars (see below) acting as beaters. (Photo courtesy of William Berry).

It had however one very fortunate advantage. At this time when Grouse Disease was becoming rampant in Scotland, Tentsmuir remained free of this infection. Figure 6.8 shows a grouse drive on Tentsmuir in progress on 21st September, 1904 with cadets from the training ship Mars acting as beaters (Figure 6.9). (The bag on this occasion was 20 brace of grouse, plus 1 partridge, 1 snipe and 2 hares (Game Book, Tayfield House - courtesy of William Berry 6th of Tayfield). For 28 years the annual total bag only fell below twenty brace three times. The population seems to have been limited by a lack of water for the young birds in dry seasons (Smith, 1948).

Grouse disease

At the beginning of the 20th century the Grouse populations of Scotland were suffering from an outbreak of a cyclical disease of unknown cause. In June 1904 a Commission was set up by the Board of Agriculture under the chairmanship of Lord Lovat (Figure 6.11) to investigate this disease. Mr. William Berry 4th of Tayfield was the secretary to the Commission. It might therefore be presumed that he had some concerns in relation to this disease affecting his new populations of Grouse at Tentsmuir. The Commission had no laboratory of its own, and only a limited budget. This consisted of provision for just one Field Observer and even this post was restricted to an engagement of only 6 months in the year, as that was considered all that was necessary.

Figure 6.9 The Mars, a retired naval vessel kept on the Tay estuary as a cadet training ship. (Photo. courtesy of William Berry).

Figure 6.10 Photograph taken around 1890 of William Berry 4th of Tayfield, the shooting tenant at Tentsmuir, sitting at a butt, with one foot reported as being in a rabbit hole. The large number of rabbit holes served as nesting refugia for Shelduck. (Photograph from Harvie-Brown -1906).

Figure 6.11 Simon Fraser 14th Lord Lovat (1908). (Source Wikipedia.)

In 1905 Lord Lovat (Figure 6.11) was introduced to Edward Wilson who had just returned the previous year from the 1901-04 Discovery Expedition with Captain Scott to Antarctica. Wilson was offered and accepted the post of Field Observer, partly because it meant that having the winter free would allow him time for his other activities in sketching and writing (Seaver, 1933).

Wilson's work as the Field Observer to the Grouse Disease Commission soon began in earnest. He recorded in his diary that, 'grouse came pouring in by every post' to the Wilson home at Bushey (Hertfordshire). It seemed that practically every dead grouse found on a British moor was dispatched to Wilson

Over the course of the Inquiry he would dissect just under 2,000 birds. It was at this time that Edward Wilson sat for the famous portrait of him by Alfred Soord (Figure 6.12) Sitting for this portrait allowed Wilson some unusual moments of quiet. Badly in need of rest, Wilson had to 'sit absolutely still for four hours on end for 3 or 4 days running.

Edward Wilson became acquainted with Tentsmuir after he became the field observer for the Grouse Disease inquiry, when he had close working association with Mr William Berry (Tayfield-Tentsmuir), the Secretary to the Board of Agriculture Grouse Commission.

Figure 6.12 Edward A. Wilson (1872-1912) by Alford Usher Soord (1868-1915), a British painter whose most famous work was a painting of *The Parable of the Lost Sheep*. Sitting for this portrait permitted Edward Wilson some unusual moments of quiet. (Reproduced with permission from Scott Polar Research Institute, Cambridge).

There is an entry in Edward Wilson's diary which marks the 5th of June 1909 as '*a red letter day in my ornithological education*' and continues '*We, that is Eagle Clarke, W. Berry, Willie Carstairs,* (the keeper), *and self spent the whole day on Tentsmuir and the seashore, approaching by Morton's Lough, we saw things on moor, heather, rushes and sand.*' He then goes on to list 26 species of birds that he had seen at Tentsmuir in the various habitats as well as numerous species of moths and butterflies (Seaver, 1937) (Table 6.1).

Table 6.1 Birds recorded by Edward Wilson as seen at Tentsmuir on June 5th 1909 (Seaver, 1937).

Curlew	Common Tern
Sparrowhawk	Lesser Tern
Carrion Crow	Sandwich Tern
Wheatear	Arctic Tern
Stock-dove	Greater and Black
Eider-ducks	-backed Gull
Coots	Peewit
Moorhen	Ringed Plover
Mallard	Golden Plover
Teal	Snipe
Linnets	Sheldrake
Grey Wagtails	Shoveler
Yellow-hammers	Redshanks

Figure 6.13 Lord Lovat's shooting Lodge near Whitebridge. The Lodge was a present from his wife to celebrate Lord Lovat's safe return from the Boer War. Edward Wilson frequently used the Lodge during his Grouse disease investigation (Old photograph courtesy South Loch Ness Heritage Group).

The ornithological richness of Tentsmuir at this time was evident in the attention it was given by a number of distinguished ornithologists. These included, William Eagle Clarke, Harvie-Brown, Edward Wilson, William Berry and others, all of whom hoped that one day they would be able to create a Reserve at Tentsmuir Point (Dr. John Berry, pers. comm.).

Wilson also undertook research into the Grouse disease further north in the Scottish Highlands. Lord Lovat (Figure 6.11) put his shooting lodge at Whitebridge at Wilson's disposal and the Wilsons frequently stayed there during the Inquiry (Figure 6.13).

The grouse disease was already known to operate in 4 year cycles. This could explain why, despite the fact that grouse were continuing to arrive in the post, none had been found to have the dreaded disease. It took 3 years of careful investigations before an outbreak of the disease occurred again. Wilson's suspicion as to the cause of grouse disease had by then started to fall upon a minute threadworm which infected the cœca of weaker birds and appeared to interrupt the digestive process, so proving fatal.

It was then, after a three year search for the alleged grouse parasite, that it was with considerable delight to Wilson, to discover the worm at last alive, and wriggling in a diseased grouse lung, to watch them hatching and burrowing in the gut, and to find dead worms in the liver. He was at last getting close to being able to prove that the threadworm was indeed the cause of the grouse disease. Wilson also traced the mode of infection to the dew drops on the tips of the young heather shoots upon which the grouse feed. He was therefore at last able to conclude that the 'grouse disease' was caused by the nematode worm

(*Trichostrongylus tenuis*), which lives in the bird's gut. A single bird could be infested with more than 30,000 worms, causing internal bleeding and reducing the number of grouse eggs. The infection is spread through the worm's eggs and in grouse droppings. When they hatch they live on the heather, which forms 90 per cent of the grouse diet. Fortunately, the Grouse population at Shanwell (Tentsmuir) proved to be completely free of the disease. This was probably due to the recent introduction of the birds and their effective means of controlling the infection (see extract below from an appendix to the report). At Tentsmuir salt flooding was facilitated at this time by the digging of ditches to encourage the access of salt water (Berry, 1939). By mid-1909 Wilson had been confirmed as the Chief of Scientific Staff for Robert Falcon Scott's (1868-1912) forthcoming expedition to Antarctica. He was now so tired that he took to working standing up so that he did not fall asleep.

Wilson was still trying to finish the grouse report when he sailed south in June 1910. The final instalments were posted home in August to the Commission from South Africa and at last his grouse work was completed (Seaver, 1933). Wilson died with Scott and never saw the final report which was published in 1911.

Bird migration

Alfred Newton, Professor of Comparative Anatomy at Cambridge, a leader in the founding of the British Ornithologists' Union persuaded in 1888, the British

---

Extract from *The Grouse in Health and in Disease* (London 1911 Vol. II Appendix I pp138-139) confirming the absence of all traces of Grouse Disease at Tentsmuir by Edward Wilson

The moor in question is in the county of Fife. A stretch of sandy soil of about 1,000 acres lying on the edge of the North Sea, and only a few feet above the high water mark – it has a good but somewhat irregular growth of heather. Until 1872 there were no Grouse on the moor, but in that year a few wild birds were turned down, and speedily became established. The moor now yields an annual bag of from forty to sixty brace. This moor is entirely free from any appearance of Strongylosis and the Grouse obtained from it are the only Grouse examined by the Committee which, on dissection, show no trace of the Strongyle worm. The absence of this parasite may be due to the fact that the moor is isolated from other Grouse ground; but this can hardly be the correct explanation, seeing that the original wild birds by which the moor was stocked must presumably have been infected with the normal quota of this nematode. A more probable explanation is that the salt from the sea spray has so impregnated the ground as to make it impossible for the worm to exist, for it has been proved by experiment that even a mild solution of salt is fatal to the Strongyle in the larval stage. On the other hand, hand-reared Grouse are often entirely free of the Strongyle worm, and it is for this reason they are the only birds which can be usefully employed for experimental purposes.

Association for the Advancement of Science to sponsor a committee for the study of bird migration throughout the world and especially along the coasts of England and Scotland. Harvie-Brown, John Cordeaux, and William Eagle Clarke were active in recruiting the assistance of the keepers of lighthouses and light vessels in recording observations of bird movements for these migration studies.

In relation to this early beginning of the study of migration, the sudden and unexpected arrival of Pallas's Sand Grouse (Figure 6.14) at Tentsmuir in the Spring of 1888 brought this coastal region of Fife to the attention of those ornithologists interested in migration. Harvie-Brown himself shot three of the these birds. The arrival of the Sand Grouse came at a time when active interest was awakening generally in the phenomenon of bird migration and brought Tentsmuir to the fore as a site for the study of bird movements.

The location on Shanwell Muir where the birds chiefly settled is marked on the 1906 map of Tentsmuir prepared by Mr. William Berry (Figure 6.7) Sand Grouse however, have only been recorded four times at Tentsmuir since 1900.

Pallas's Sand Grouse is an irregular visitor and 1888 appears to have been quite exceptional, as large numbers of these birds also appeared widely throughout western Europe. It was estimated that overall 2000 birds reached Scotland (Baxter and Rintoul, 1953). There was apparently at this time a mass migration of Sand Grouse across a large part of north western Europe from their very extensive home range extending from the Steppes of Central Europe and as far east as China.

Dr. Eagle Clarke, who took an interest in the arrival of the Sand Grouse, is still remembered as having had a significant influence in laying the foundations of modern knowledge of bird migration, and was a leading exponent in Britain of new methods in ornithology. In particular he promoted the study of the routes that birds followed in their migration. His work stimulated fresh collections and study of records in Europe, and when in 1903 the British Association for the Advancement of Science formed a Bird Migration Committee, Eagle Clarke was chosen to prepare the final reports, the last of which was presented to the Southport meeting in 1903. In earlier days, intensive observation of the movements of birds was limited to specific localities. Now, new efforts to co-ordinate observations from

Figure 6.14 Pallas's Sand Grouse from a colour lithograph in the second edition of Johann Friedrich Neumann's classic *Naturgeschichte der Vögel Mitteleuropas*, published by Gera-Untermhaus in 1896-1905.

Figure 6.15 Map of North Tentsmuir and the Tay Estuary showing in red the land bought by the Town Council of Dundee with a view to establishing a new local industrial area for the prefabrication of ship parts together with an onshore welding assembly facility for ships to be finished in the Dundee yards. (Map by courtesy of the Archives of the City of Dundee).

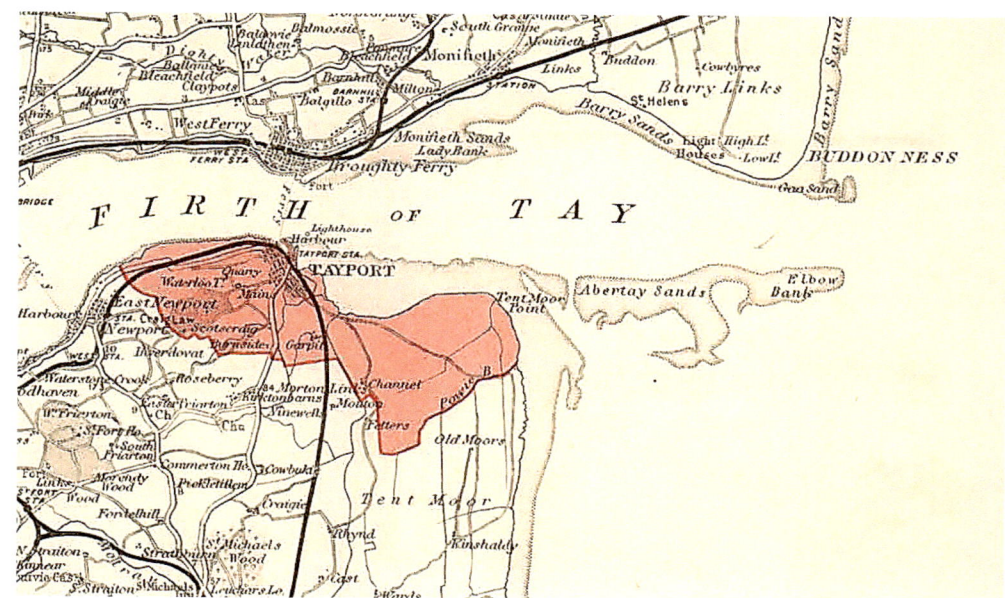

many areas began to provide the information upon which interpretation of migratory journeys depended.

At this time bird-ringing had not yet been tested as a scientific method. This had to wait until the Danish ornithologist Hans Christian Cornelius Mortensen (1856 –1921) developed the technique of ringing birds for scientific purposes in 1899. He published reports regularly with Dr. J.A. Harvie-Brown and other local ornithologists. Dr. Eagle Clarke remained a regular visitor to Tentsmuir studying bird movements until his death in 1938.

20th century Tentsmuir

During the 1914-1918 war the pine woods bordering the moor were cut down. For the nesting birds this was beneficial as it removed the crows that had been in the habit of marauding their nests.

In 1919 the estate of Scotscraig was sold to Dundee Town Council by the Maitland Dougall family who had owned it for most of the previous century. This purchase was made with a view to establishing a new local industrial area (Figure 6.15). This area was to be used for the prefabrication of ship parts together with an onshore welding assembly facility for ships which could then be completed and fitted out in the Dundee yards.

It appeared that the purchase was made principally to deter a possible competitive bid from a Belfast shipyard. No such development ever took place and for a short time a warden was appointed to safeguard the birds during the nesting season.

The afforestation of Tentsmuir

On 1 September 1919 the Forestry Act came into force with responsibility for woods in England, Scotland, Wales and Ireland. In 1921 the Forestry Commission made its first land acquisition at Tentsmuir by purchasing the central part of the Kinshaldy Estate where it immediately started planting trees (Figure 6.16). This has been described 'as bringing to an end some fifty years of active protection of ground-nesting birds on an area which had long been noted for its populations of birds of great interest and ecological value' (Grierson, 1962).

In 1924 the land that had been bought by Dundee City Council for ship building, but never developed, was also purchased by the Government on behalf of the Forestry Commission. Tree planting then took

place in earnest at the north end of Tentsmuir. Much of the of the hinterland behind the dunes was planted with conifers and the moorland rapidly vanished with severe consequences for the remaining ground nesting shore birds and waders.

It was however, not just the planting of trees that caused the most drastic change in the moorland bird populations. As pointed out by Dr. Berry, they could have nested in the shelter of the young trees. It was the drainage, and above all the erection of narrow mesh fencing which prevented the young birds from reaching the water and feeding. This resulted in these young birds dying in their thousands and immediately stopped Tentsmuir being a suitable breeding area for moorland and coastal birds. These changes were very clearly pointed out in the account of the *Status and Distribution of Wild Geese and Wild Ducks in Scotland* undertaken by Dr. John Berry of Tayfield in collaboration with Misses Baxter and Rintoul (Berry, 1939).

Another environmental-changing action by the Forestry Commission at Tentsmuir was to drain the wet moorland that had for long provided ideal nesting ground for both ducks and other birds. The nature and intensity of the Forestry Commission drainage operation and fencing can be seen in photographs and maps of that time (Figures 6.16-18).

Another unforeseen consequence for the birds nesting on Tentsmuir arose from the energetic trapping and blocking up of rabbit burrows in the area. The expulsion of the rabbits inadvertently contributed to the disappearance of the Shelduck which were used to nesting in the rabbit holes (Berry, 1939).

Yet another and more serious consequence of the removal of the rabbits was not anticipated, but soon became evident. The foxes which previously had paid only occasional visits to the moor and did little or no damage to the population of Grouse and other ground-nesting birds, as there was a plentiful supply of rabbits at their disposal elsewhere, now turned their attention on the ground-nesting birds.

By 1928 the fox population had increased enormously and found the new plantations to be an excellent sanctuary. This problem continued and by 1948 the fox population had become a serious menace to all ground nesting birds, not only on the moor but for miles around. (Dr John Berry: pers.comm).

The initial planting of conifers was over an area of about two square miles on land enclosed by a fence that ran for several miles close to the sea around the estuaries of the Tay and Eden (Figure 6.18). As Dr. Berry (see frontispiece) pointed out in his seminal work on the *Wild Geese and Wild Duck in Scotland* (1939)

Figure 6.16 Photograph taken shortly after the acquisition of Tentsmuir Point by the Forestry Commission showing the drainage that had been put into Tentsmuir point for the conifer planting by the Forestry Commission (Photo reproduced with permission from Berry, 1939).

Land, people and resources 97

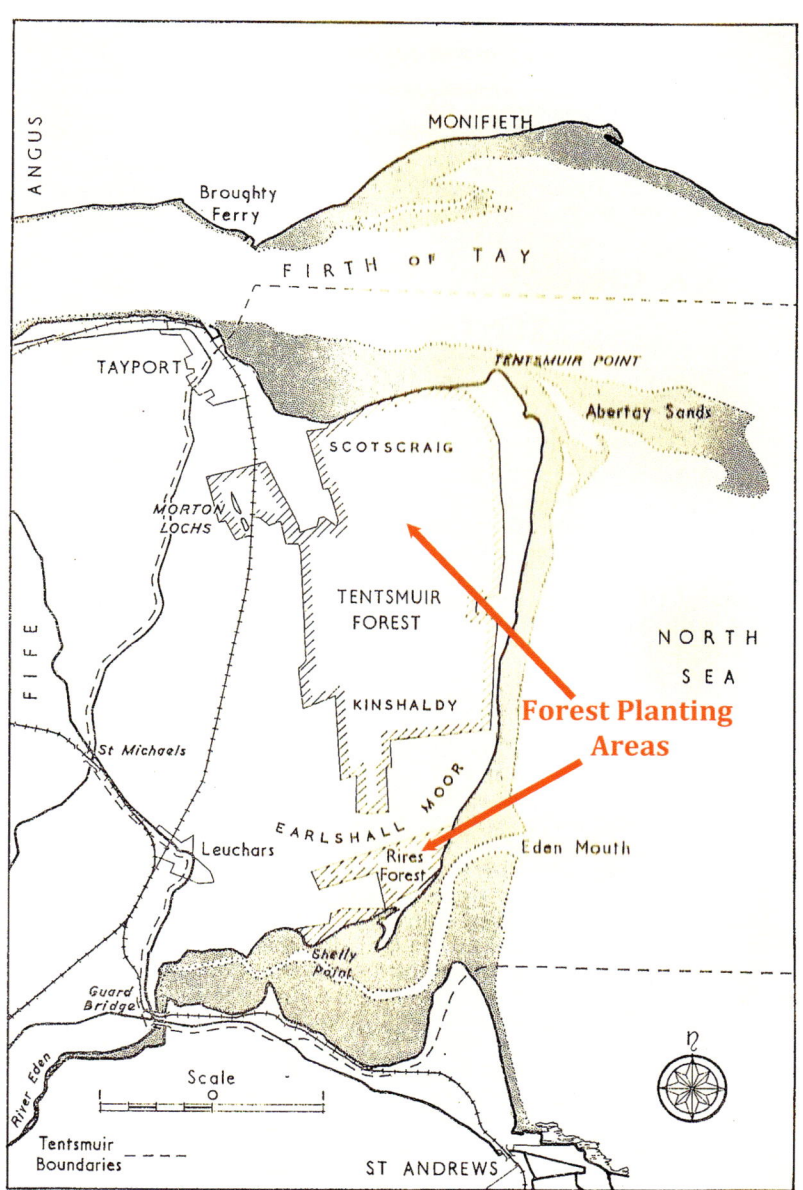

Figure 6.17 Areas planted by the Forestry Commission beginning with the central Kinshaldy area in 1921 and then followed in 1924 by the northern portion at Scotscraig (Adapted map reproduced with permission from Grierson, 1962).

'if this land had not been enclosed in this way it might still have provided an attractive breeding ground for several species of ducks.' However, the fence was placed close to the foreshore and used a mesh which effectively prevented the ducklings from reaching the shore for water and feeding. Large numbers of young Eider and Shelduck and other species were therefore killed in this way.

Dr. Berry also noted that as the trees began to grow and more of the area was planted with trees, the presence of heather was much reduced and was replaced by thick grass which previously the rabbits had kept in check. Gradually, the nature of the vegetation changed and the supply of water disappeared, except during unusually wet seasons, due to drainage and afforestation.

When, in 1949, the National Parks and Access to the Countryside Act was passed and the Scottish Nature Conservancy was created, it was to the advantage of the wild life of Tentsmuir that Dr. Berry was appointed as its first Director. His appointment was in fact made before the Scottish Nature Conservancy itself had come into being. The Secretary of State for Scotland (Joseph Westwood) had been determined not to be outflanked by

Figure 6.18 Photograph taken some time before 1939 showing the conifer planting by the Forestry Commission and the proximity of the enclosing fence line to the shore (Photo reproduced with permission from Berry, 1939).

London and made a pre-emptive strike to secure Dr. Berry to lead the new Nature Conservancy (Figure 6.19)

In 1954 Dr. Berry (see frontispiece) arranged the purchase of Tentsmuir Point from the Forestry Commission for the sum of £120. He had explored this area since he was a very young boy and had long appreciated the ecological importance of its many species of birds and plants In his notes from that time he gave an account of this purchase. He recorded that the only area that could be obtained for the Nature Conservancy as a reserve was the strip that had been below the high water mark at the time when the Forestry Commission acquired all the land they now possessed. Dr. Berry had observed however, that this strip was now accreting and he felt confident that this would continue, and also that it might be expected that the accretion would be fast. This proved to be a very perceptive prediction and the result was a most felicitous purchase in terms of land accretion.

Although there has been significant erosion in recent years the reserve is still more than three times the size it was at the time of purchase in 1954.

Tentsmuir drainage

Drainage at Tentsmuir has long been an obstacle both for farming and forestry. The natural tendency of the accumulated sand to impede drainage created the moisture laden wetlands and moorlands that have

Land, people and resources

Figure 6.19 Morton Lochs looking north. With the South and northern Lochs in the centre of the image and the small West Loch to the left. Note the line of the 1849-1959 railway line to the east the remains of which now serve as a drainage barrier for water flowing to the sea. Aerial photograph taken in 1952 when the Morton Lochs region first became a nature reserve (see text).

provided both nesting and over-wintering habitats particularly for wildfowl. Already in the 17th century the need for drainage was recognised and the extent of the wetlands was clearly illustrated in Bleau's Atlas of 1654 (see Chapter 2).

The First Statistical Account of Scotland (1791) records the drainage activities of the farmers at Tentsmuir who at that time had to undertake extensive excavations to facilitate the exit of water through their drains. These were described at that time as long, beautifully constructed canals. One of the oldest is that known today as the Canal Loch (Ordnance Survey GB grid reference NO485220). This name has a long history which was never due to any connection with a canal. Its original name in Gaelic was the White Loch. *Loch can – can* denoting white in Gaelic (see Place Names of Fife Vol. 4)). It is further wrongly named in the Bleau Atlas of 1684 as Candle Lochs (see Figure 2.12). Other ancient lochs at Tentsmuir are the White-myre, Toremont (originally Foremont, see also Place Names of Fife- PNF) and the Tentsmuir or Big Waters.

The *Great Drain*, (see also Chapter 2) was dug by Sir David Carnegie in 1785 before he sold the lands to the hon. Robert Lindsay. The drain was three miles long and ran parallel to what is now the disused railway line (see below). Some 20 years later Robert Lindsay observed the need for an increase in the size of this *Great Drain* and had it re-excavated to a width of 20 feet and a depth of 14 feet (see PNF).

Despite the continued presence and maintenance of this major drain, the farming community at Tentsmuir have long been acutely aware of the need for regular drainage maintenance. Widespread in the soil profile throughout Tentsmuir is a layer of white sand at a depth below the soil tillage zone which readily infiltrates into the drains and leads to blockage Some farmers have therefore been in the habit of placing a suitable form of shuttering either side of their drainage pipes in order to reduce this rapid blocking of their drains. (pers.comm. Colin Mitchell, Leuchars Castle Farm).

In 1845 the Edinburgh and Northern Railway Company (ENRC) received Royal Assent to construct a railway line which was routed through Tentsmuir.

Figure 6.20 Boating at Morton 1906.

The objective was to have a railway that ran between Burntisland and Ferry-Port-on Craig (renamed Tayport by the railway company) thus opening a passenger and freight route via train and ferries to access Fife and the north of Scotland).

From the very beginning, the railway surveyors noted that the area at Morton was very wet and they needed to drain it with ditches and culverts on both sides of the line. These are still in existence today, and although silted up they still have some flow, albeit reduced. The Edinburgh and Northern Railway Company (ENRC) also excavated a large ditch on the west side, parallel to the railway line. Drainage ditches and culverts and probably soak-aways from the large ditch and wet areas (rises) were built and ran under the railway line so that the water flowed into a system of ditches eastward into the Lundin Burn. Water also flowed from the railway drainage system north and eventually flowed into the River Tay via local Tayport burns. The railway track is still a prominent feature in the Tentsmuir landscape (Figure 6.19).

Figure 6.21 James Christie of Scotscraig and daughters (extreme right in photo) stocking Morton Lochs 1906 (Photo SNH archive).

The Tayport railway line was closed to passenger service in 1956 and finally closed to freight traffic in 1959. Long after the closure of the railway line, the remains of its embankment and ill-maintained culverts have continued to be a hindrance to natural drainage as little attention has been given to the maintenance of the drains that are needed to prevent the former

Land, people and resources 101

Figure 6.22 Whooper Swans, a species that is regularly seen at Morton Lochs (Photo Author).

Figure 6.23 Goldeneye on Morton Loch. ((Photo Lorne Gill - SNH).

Figure 6.24 Little Grebe on Morton Lochs (Photo Professor Stephen Buckland-SNH).

collapsing railway embankments from blocking the much needed culverts.

Given the many impediments to drainage, with so many open dykes it is no wonder that Tentsmuir, which used to be known as Sheughy Dyke (Wet ditches), still has serious drainage problems.

Morton Lochs

Impediments to drainage at this northern part of Tentsmuir probably suggested to the Christie family, who farmed land at Scotscraig, including an area called The Rises and its extensive wetlands and flooded slacks, that better use could be made of the wetlands to the west of the Railway Line by encouraging rather than trying to defeat the flooding. They therefore decided in 1906 to create large fish ponds by excavating around the existing wetlands and diverting a convenient stream, the Ninewells Burn (now called the Lead Burn) to create the three Morton Lochs – North, West and South (Figures 6.20-22). Due to the flooding that already existed in this area, the creation of these ponds did not need a total excavation, but merely some deepening. The Lochs were initially stocked with Carp imported from Italy as well as Brown Trout, Eels, Minnows, and Sticklebacks. Gudgeon were also introduced for sale. Thereafter, the lochs soon became an important centre for wintering wildfowl.

Creation of Morton Lochs Nature Reserve

In May 1952 Morton Lochs were also declared to be a National Nature Reserve when the lochs with surrounding land to the extent of 47 acres was purchased from the Forestry Commission. In 1956 a further 12 acres were added to the Reserve giving a total area of 59 acres (24 hectares).

From an ornithological point of view, these open fresh water lochs are of considerable significance as they lie at the junction of two important migration routes. One follows the coast southwards from N.E. Scotland while the other, turns inland and up the Tay and thence via Loch Leven and the upper Firth of Forth to the Scottish West Coast and Ireland. There are no other fresh water lochs within 8 miles hence their importance for large numbers of waterfowl and waders which visit here on migration. From winter and through to spring the visiting swan and duck species include, Shelduck, Pintail, Teal, Mallard, Gadwall, Wigeon, Shoveler, Pochard, Scaup, Whooper Swan, Goldeneye and Little Grebe (Figures 6.22-4).

Unfortunately, the silt load in the burns feeding Morton Lochs is such that the Lochs can readily dry up. In the 1960s the area of open water was much reduced due to advancing vegetation which included Water Lilies and Iris (flags), Bulrushes, Horsetails, with Bracken on the drier ground surrounding the lochs. By controlling the water levels some recovery was achieved and Grebes returned to the lochs to breed.

When Tentsmuir became a nature reserve in 1954 Dr. Berry had wanted to include Morton Lochs in the one reserve. Initially, this plan was resisted, and the lochs were allowed to become largely overgrown with grass and reeds. In addition most of the surrounding land was planted with trees by the Forestry Commission which made the access unsuitable for larger wildfowl (Figures 6.23-25).

Figure 6.25 Detail of swamp vegetation in 1967 largely dominated by common reeds (*Phragmites australis*). (Photo SNH archives).

Fortunately this has now been radically changed and the lochs are now a National Nature Reserve and managed in conjunction with the Tentsmuir Reserve. The trees planted by the Forestry Commission have also been removed to facilitate the access of birds to the open water.

By 1970 the lochs had dried out several times due to a build up of silt and wind-blown sand. A decision was then made to re-establish the conservation value of these lochs and improve their design by constructing islands and promontories in order to provide improved nesting and roosting sites.

In July 1976 restoration work began, first by closing the inflow sluices and drying out the lochs so that they

Figure 6.26 Aerial view looking south over North and South Morton Lochs over-grown and choked with vegetation in 1967 (Photo SNH archives).

Figures 6.27 (a-c) Stages in restoration of the North Morton Loch in 1976 (a) Entire loch cleared of vegetation. (b) Old tyres used to make foundations for new islands. (c) Islands and promontories formed to increase shorelines and improve nesting conditions for wildfowl. (Photos SNH archives).

could be excavated. This proved to be a considerable task and specialized pumping equipment had to be brought in to complete this stage in the restoration. There were frequent surprises, one of which was the presence of some 9,000 eels.

a

b

c

Once the lochs had been completely excavated artificial islands and promontories were created in such a manner as to extend the shore line. In this way both nesting and roosting for the wildfowl were greatly increased. Care was also taken that a sufficient flight path was left for the larger birds to be able to land and take off from open water. Figures 6.25-28 show these stages in reconstruction and the final result as seen from the air.

Tentsmuir forestry

Small shelter belts along with larger strips of forest began to be planted at Tentsmuir in the 19th century, particularly around Kinshaldy and Fetterdale. However, it was not until the Forestry Commission came into being on the 1st September 1919 that planting of State Forests began at Kinshaldy in 1921. By 1924 planting on a large scale in central Tentsmuir was underway with extensive draining and fencing. This rapidly extended both north and westwards until only a small strip of land was left on duneland on the north coast. The initial planting was of Scots Pine (*Pinus sylvestris*) from East Anglia and Corsican Pine

(*Pinus nigra*). In these early years the Corsican Pine grew well on the dry sandy soil, producing little seed but developing tall cylindrical boles (Figure 6.30). The Scots Pine grew less well and had a lower yield than

Figure 6.28 Oblique aerial view looking south the restored North Loch and the South of Morton Loch in the distance (Photo SNH archives).

Figure 6.29 View of the restored North Loch showing the islands and the preserved open water sufficient for the landing and take-off of larger birds (Photo Author).

Figure 6.30 Corsican Pine (*Pinus nigra*) with well-formed tall trunks despite wind exposure on the dunes bordering the south bank of the river Tay (Photo Author 2016).

would have been expected. The Scots Pine seeded prolifically, a characteristic that later caused the nature reserve to be invaded with pine when myxomatosis devastated the rabbit population in the early 1960s (see Chapter 10).

As a result of the poor performance of Scots Pine, the Forestry Commission then started to plant Sitka Spruce (*Picea sitchensis*). However, the denser nature of the Spruce plantations and the resulting litter, sadly had a negative effect on the ground flora which had managed to survive under the pine plantations.

Water table levels

A study of the effects of afforestation on soil and water table levels was carried out around 1950 on two areas that had been planted with pines for 11 - 19 years respectively (Figure 6.31-32) in a region where the water table was generally close to the surface (Ovington, 1951). The water tables rose in winter and were generally higher in the more inland areas where drainage seawards was impeded by restricted drainage.

Ovington considered that the planting of the trees lowered the level of the water table. However, in more recent times the marked advance seawards of the front line of dunes and the ever present tendency of the seaward drains to become blocked has tended to raise the level of the water table. This is most marked in cold weather in winter when the transpiration of the forest and surrounding vegetation is reduced.

These profiles indicate some drying of the soil in the initial stages of forest development. When this study was carried out the forest canopy was described by Ovington as 'just beginning to close'. In Figure 6.32 the lowest soil layers are described as white sand. This matches the description given by local farmers at that time who were used to having problems in maintaining drains that readily blocked the mobility of the silver sand and the consequent need for additional shuttering to prevent the drains from becoming quickly ineffective.

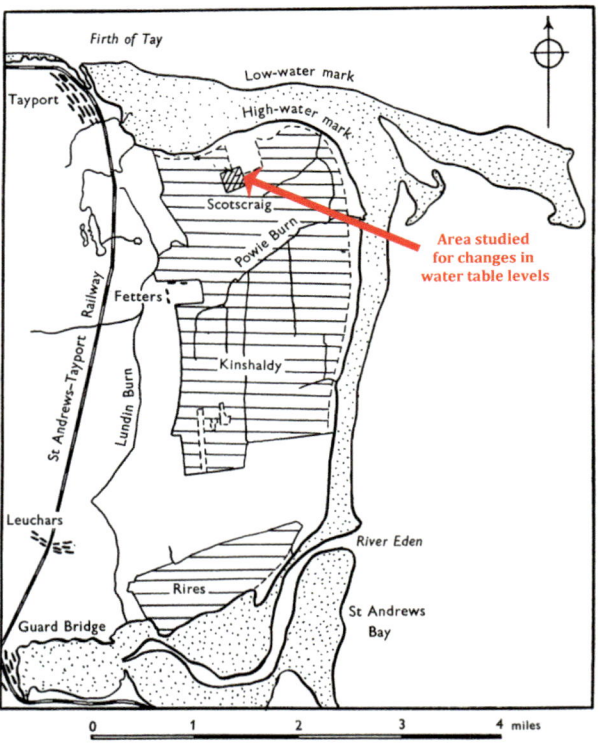

Figure 6.31 Map of Tentsmuir Forest indicating the area studied for changes in water table levels and soil water content as a result of pine planting (*Pinus sylvestris*, *P. nigra* and *P. nigra laricio*. (Reproduced with permission from Ovington 1951).

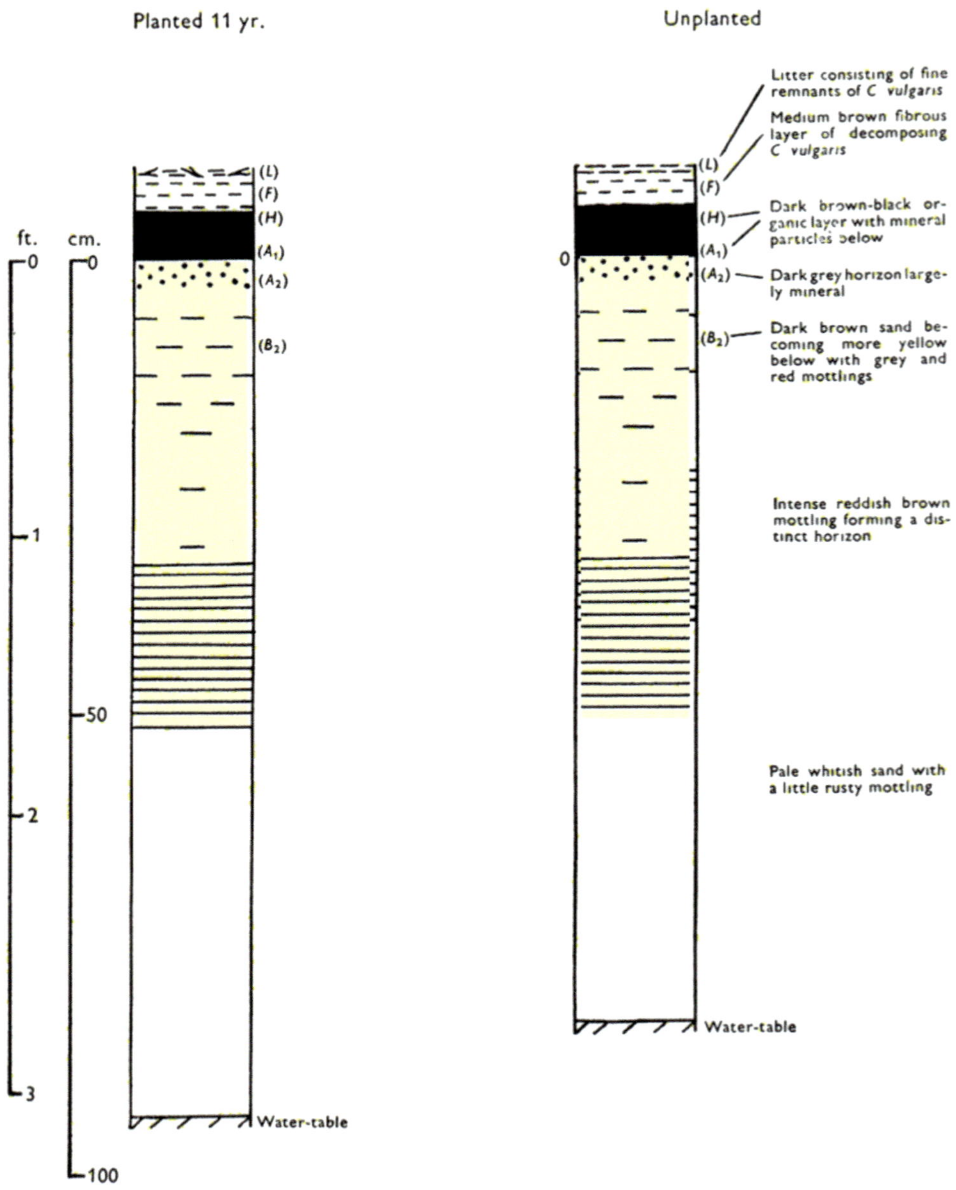

Figure 6.32 Soil profiles in study area. (Reproduced with permission from Ovington 1951).

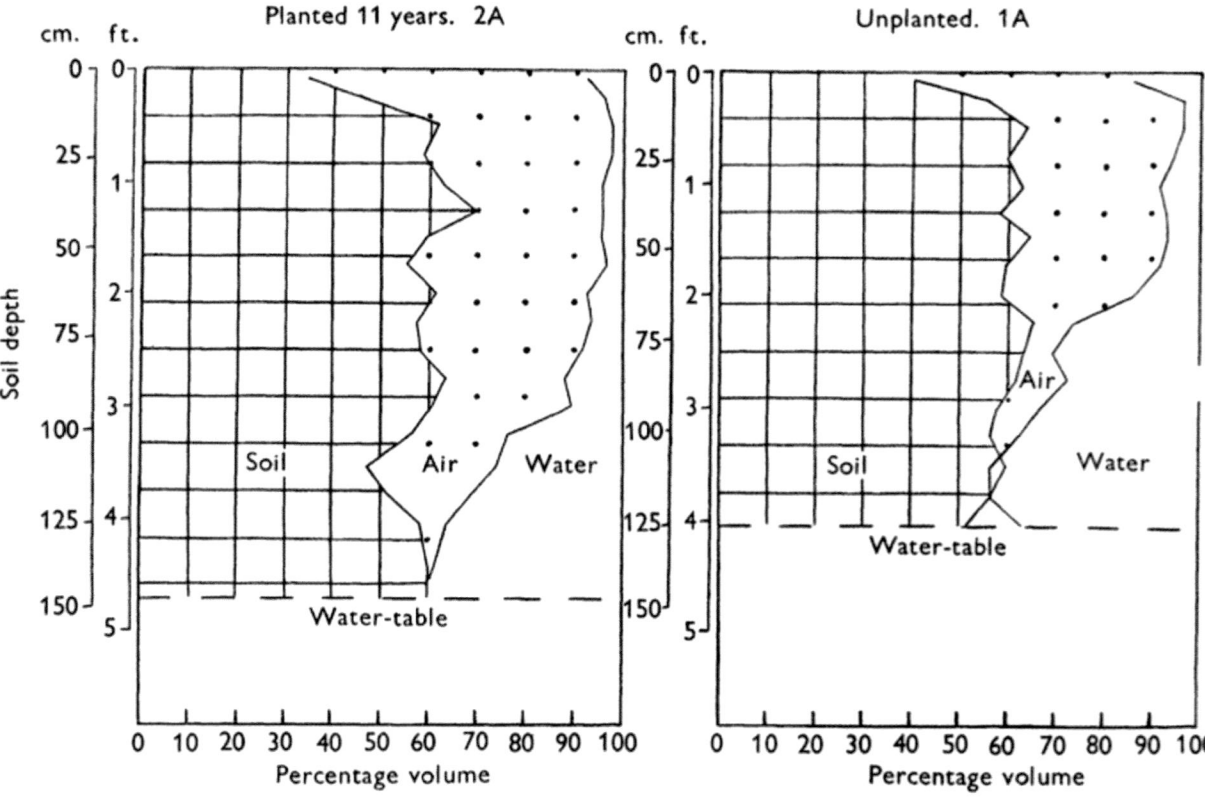

Figure 6.33. Percentage volumes of soil, water and air in the soil profiles (Reproduced with permission from Ovington 1951).

Within the forest there is a tendency for the height of the water table in summer to be too low for species such as Sitka Spruce, and too high in winter for Scots and Corsican Pine. There have been many attempts to remedy this situation by improving drainage but this has become more difficult with the rapid growth and expansion of the sand dunes to seaward that took place as the forest was initially developing.

One solution that was reluctantly agreed to by The Nature Conservancy was to create a by-pass of the ancient and historic Powie Burn (see Chapter 2). A deep ditch was therefore dug from where the Powie Burn enters thereserve and ran south to take the water that would have entered the slacks to a drainage pond situated outside the reserve. This was described as the *Powie Burn By-Pass* (Figure 6.34). As well as providing an effective forest drain, it was hoped that the forest pond might be attractive for waterfowl. However, this ditch quickly became filled with silt and blown sand and has proved so ineffective that attempts to keep the drains clear have long been abandoned (Figure 6.34).

Figure 6.34 The Powie Burn By-Pass Pond shortly after its construction in March 1968 (Photo Malcolm Smith).

Figure 6.35 One example of windthrow, which is widespread throughout the forest (Photo Author 2016).

Examination of the state of the plantation at present shows a series of contrasting sites. To the north on the substantial and elevated dunes near the mouth of the Tay, Corsican Pines stand erect with well-formed trunks despite exposure to the shore winds (Figure 6.30). To landward however, where the water tables rises every winter there are areas that are suffering from extensive windthrow (Figure 6.35).

Although the market for good well-formed timber has disappeared, as there is no modern demand either for telegraph poles or pit props, there is nevertheless a growing demand for timber as a biofuel. In this respect Tentsmuir is well placed in proximity to this market. Unfortunately, fresh planting of Corsican Pine is at present not allowed due to the presence of widespread infection of *Dothistroma* Needle Blight (DNB).

Land, people and resources

Figure 6.36 Future projection of age stands of the principal species at Tentsmuir Forest (Figure from Forestry Planning Document (2016 courtesy of R. Lofthouse).

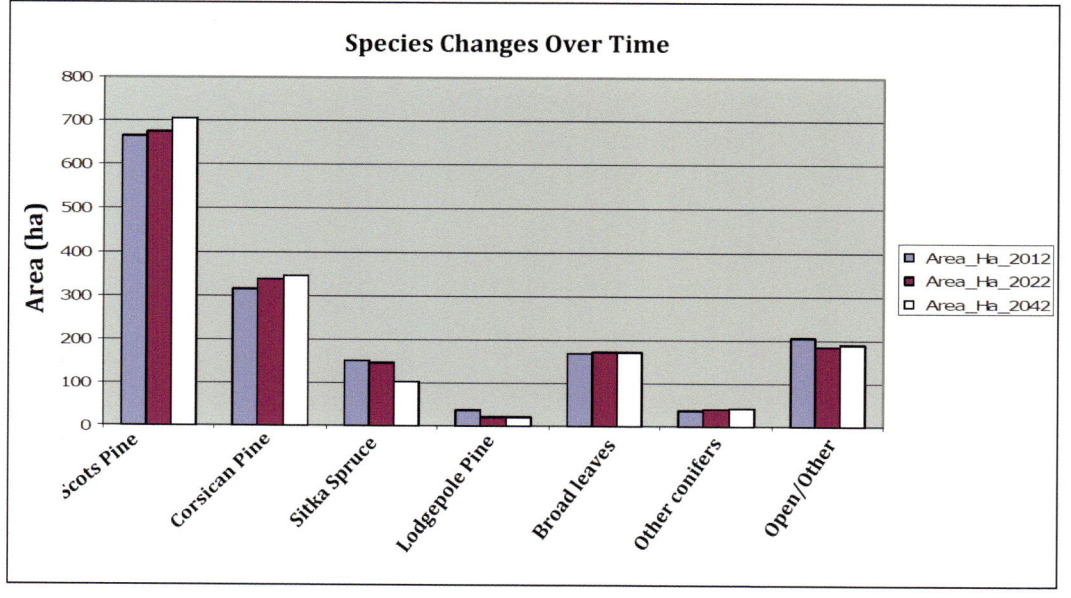

Figure 6.37 Present and envisaged future species composition of forestry tree species at Tentsmuir. (Forestry Commission, 2016).

This fungus was first recorded in the forest in 2005 on Corsican Pine and since then almost all pine stands have been affected, including Scots Pine which was initially thought to be resistant. Younger stands appear to be the most affected. Young trees have lost one to two years of needles and growth has practically stopped. A new thinning regime has been implemented, following guidance from Forest Research, to remove around 50% of standing trees in order to decrease humidity levels under the canopy and decrease infected material. Pruning is also being considered to improve timber quality.

The forest is therefore now operating as far as possible a policy of *continuous cover forestry* which started in 2002. There are however serious difficulties in implementing such a policy given the extent of the fungal infections.

Visitors to the forest and adjacent reserve are a significant benefit to the Forestry Commission in terms of the revenue that is gained from the provision of car parking and other visitor facilities. The forest is also an environmental asset to the area as a whole and plays an important role jointly with the nature reserve in preserving biodiversity in the region. There are many diverse areas which require a joint strategy between the Commission and the Nature Reserve that could be a benefit to both. One example is the management policy now adopted by the Forestry Commission to avoid clear felling where possible and to produce a forest with a greater variation in age classes, more mature trees, and a richer ground flora (Figures 6.36-37).

Areas of joint interest in relation to conservation of Tentsmuir as a whole are discussed in Chapter 10. Recent attractions, such as White-tailed Eagles and Ospreys, are all made possible by a partnership between forestry and conservation.

Chapter Seven

Tentsmuir's thriving birds

Bird habitats at Tentsmuir

Tentsmuir's coastal situation between two major estuaries is outstanding for the wide range of habitats it provides for bird life. Within a relatively restricted area of a 7 mile coastline and a 2-3 mile wide hinterland, Tentsmuir contains sandy shores, salt marshes, heathlands conifer plantations and lochs. Each of these habitats is frequented by its own characteristic bird species. The mildness of the environment throughout the year, and the diverse feeding possibilities of the adjacent estuaries of the Tay and the Eden, provide both summer and winter locations for large numbers of wildfowl and geese. In autumn large numbers of geese migrate here from Greenland, Iceland, Scandinavia, Spitsbergen, Northern Russia, and as far away as Siberia. In addition, numerous species are recorded every year on passage during the migration period. On winter evenings the size of the flocks of geese attract particular attention, both visually and orally, as they arrive to roost (Figure 7.1).

Winter visitors

The number and variety of birds visiting Tentsmuir, as elsewhere in the British Isles, is influenced by the East Atlantic Flyway (Figure 7.2). This Flyway is a broad migration route used annually by over 90 million birds, as they pass from their breeding areas in North America, Greenland, Iceland, Siberia and Northern Europe to winter in the more southern and western regions of Europe and even as far as Southern Africa (Figure 7.2).

Migrating birds tend to follow great circle routes. Although these are the shortest routes, they can entail crossing physically challenging mountains and

Figure 7.1 Pink-footed Geese flocking on the Eden Estuary (Photo Professor Stephen Buckland).

considerable expanses of ocean, particularly for birds that nest in arctic regions. Rest stages are therefore necessary, and Tentsmuir is well-placed to serve as a stopover site for many birds.

The advantage of British coasts such as Tentsmuir for over-wintering wildfowl, becomes apparent when winter mortality rates for birds that use the eastern shores of the North Sea are compared with those on British shores. In Sørlandet (Southern Norway), the extent and duration of freezing, even on salt-water habitats, can be such that when harsh winters take hold, mass starvation can occur among the over-wintering wildfowl. The death toll can be enormous. In March 1986, freezing temperatures caused the death of 18,000 water birds, including 14,000 diving ducks. A similar event off the Danish coast, in the winter of 1981-82 brought about an Eider death toll estimated at 21,000 (Wrånes, 1988). Even in the relatively mild winter of 1999-2000, over 21,000 Eiders died out of a total population of 120,000 (Camphuysen et al., 2002). Dissected birds in the Wadden Sea were found to be severely emaciated. The high level of winter mortality appeared to be aggravated by a shortage of mussels and cockles.

Figure 7.2 Map of the East Atlantic Flyway showing the breeding range of arctic breeding birds. (Reproduced with permission of the Wadden Sea Flyway Initiative and Bird Life International).

Table 7.1 Species of geese and swans recorded at Tentsmuir, including Morton Lochs, and the offshore regions of St Andrews Bay and the neighbouring Tay and Eden Estuaries (Data collected by local reserve managers).

Geese	**Swans**
Pink-footed Goose	Mute Swan
Greylag Goose	Whooper Swan
Barnacle Goose	
Brent Goose	
Canada Goose (rare)	
Bean Goose	
White-fronted Goose	

Tentsmuir's geographical situation and favourable winter climate for winter feeding, makes it a particularly attractive location for wintering birds. To ornithologists, it has also long been known as a site that provides opportunities to view passing migrants in spring and autumn. A recent complete list can be found in (Elkins et al., 2016).

Tentsmuir's geese

In winter, geese are dominant features of the Tentsmuir coastline, particularly in the evening when they return from inland grazings to roost in the security of the off-shore sandbanks. The geese which are now so predominant, namely Greylag and Pink-footed Geese have not always been so numerous.

Towards the end of the 19th century, when considerable bags of geese were recorded in the game books at Tentsmuir, the most common species was the Bean Goose (*Anser fabalis*). This was the dominant large game-goose at this time throughout northern Europe and consisted of two sub-species - the Taiga Bean Goose (*Anser fabalis fabalis*) and the Tundra Bean Goose (*Anser fabalis rossicus*). The sub-species are not readily differentiated.

On the Tay, the Bean Goose was once considered ubiquitous with no serious rival (Berry, 1939) but by 1885 a decline had sent in. When viewed globally, the Scottish populations of Bean Goose were not alone in their decline. It appears that the western Palearctic populations of White-fronted and Bean Geese as a whole, decreased dramatically from the middle of the 19th century (Baxter and Rintoul, 1953a). A few sightings are still seen in Fife of the Bean Goose, mostly of the Taiga Bean form (Elkins et al., 2016).

Pink-footed Goose (Anser brachyrhynchus)

In terms of numbers, the thousands of Pink-footed Geese which now arrive regularly at Tentsmuir in autumn (Figure 7.1) are a relatively recent phenomenon. It is not possible to say exactly when Pink-footed Geese were first noticed on the Firth of Tay. However, it was not until 1872 that Mr John Berry of Tayfield recognised a specimen among the considerable bag of geese shot by him in East Fife (Berry, 1939).

Throughout the 19th century, the Bean Goose probably remained the dominant migratory goose species in Fife. However, by the beginning of the 20th century Pink-footed Geese were wintering regularly near the mouth of the Tay in considerable numbers. These wintering populations continued to increase until 1915 when aerodromes were opened at Montrose and Leuchars and a marked decline in goose numbers was noted (Berry, 1939). The disturbance caused by the aircraft at this time would have been aggravated by the RAF pilots who were observed chasing and attempting to shoot the geese from the air. However, this practice was soon stopped (Berry, 1930).

When the RAF fighter base at Leuchars was closed in the winter of 2015-16, the number of geese roosting on the Tentsmuir foreshore alone had risen to 3000 birds (SNH records). The following year the maximum number of Pink-footed Geese rose yet again. Recently, the roosting area has extended to include the southern Tentsmuir sandbars and foreshore. Pink-footed Geese are now observed roosting and feeding over an ever wider area, including the Great Slack, and the fields surrounding Fetterdale and Morton Lochs. Tentsmuir and its associated estuaries can now claim to be a site

Figure 7.3 Gaggle of Greylag Geese overwintering at Tentsmuir (Photo Lorne Gill SNH).

of national interest for this species as over 1% of the UK population can be found roosting there (SNH records).

Greylag Geese (Anser anser)

Greylag Geese ceased to breed in England in the 18th century but still maintained small nesting colonies in Highland Scotland into the mid 20th century (Baxter and Rintoul, 1953b).

Dr. Berry noted that Greylag Geese (Figures 7.3-4) had long ceased to breed regularly in the area (Berry, 1939). Nevertheless, he observed that small numbers still spent the summer in the vast reed beds of the Tay Estuary between Perth and Newburgh, where they sometimes had even been found nesting. However, the goslings were invariably shot in August before they were able to fly. In 1928 a 'sportsman' boasted in the local press of having shot a goose and five goslings on August 1st. That date used to be an annual day of slaughter for 'goose flappers' on all the public marshes round the Firth of Tay (Berry, 1939).

Figure 7.4 Greylag Goose (Photo Lorne Gill - SNH).

Figure 7.5 Migration routes of Barnacle Geese from the Solway Firth via Norway to Spitsbergen and from Islay to Greenland via Iceland.

Figure 7.6 Barnacle Goose on a cliff nest site in N.E. Greenland. This bird, as well as all those that nested in Greenland will migrate in autumn to Islay (Photo Dr. Helge Körner).

Dr. Berry added that *'with suitable protection, the Greylag Geese might become re-established as breeding birds in the Tay area.'* He also noted that from 1920 to 1930 Greylag Geese were so abundant in winter on both sides of the Tay that it was not uncommon for a score to be shot over decoys in a morning. This breeding activity in the more upstream marshes of the river Tay provided only a small temporary improvement in their breeding numbers. About 30 pairs presently nest in Fife (Elkins et al., 2016b). However, it is possible that they have originated from a feral population. The large number of visitors now frequenting Tentsmuir and adjacent areas make it unlikely that Tentsmuir will ever be anything other than a place for grazing and roosting geese.

Barnacle Geese (Branta leucopsis)

In September or early October Barnacle Geese (Figures 7.5-6) are some of the first southward-bound migrating geese to be seen in the Tentsmuir region. The birds that are seen in Fife have mainly nested in Spitsbergen (Svalbard) and pass through Tentsmuir on their southward migration to winter in the Solway. Some isolated individuals can be seen mingling in small numbers with the Pink-footed Geese in the Tentsmuir region before they continue southwards to spend the winter on the Solway.

Other wintering geese

There are numerous records of other species of geese which have been seen occasionally in winter in the Tay. Notable among the minor goose populations are the White-fronted Geese (*Anser albifrons* - Figure 7.7). Already by 1930, Dr. Berry had noted that the European White-fronted Geese found in eastern Scotland differed from the Greenlandic birds found in the west of Scotland, by being generally smaller, with paler plumage and with bills that were invariably flesh pink, whereas the West Coast birds had beaks that were entirely yellow. He did not consider these East Coast White-fronted Geese rare, and on one occasion he was able to count 800 birds during their spring migration (Berry, 1939). In more recent times both these forms have been seen during winter in the Tentsmuir area, but only in very small numbers (Elkins et al., 2003).

Brent Geese (*Branta bernicla*) are the smallest and most maritime of winter visiting geese to be seen at Tentsmuir. They exist as two sub-species, namely a pale-bellied form (ssp *hrota*) and a dark bellied form (*bernicla*). The latter occur only as occasional visitors and the former in very small numbers wintering at the mouth of the Eden Estuary (Elkins et al., 2003). Like the Barnacle Goose they usually remain only for a short period.

Whooper Swans (*Cygnus cygnus*)

Whooper Swans, like arctic geese, migrate south in autumn and many spend the winter in the Tentsmuir region feeding by day in the neighbouring fields (Figure 7.8), and roosting at night, in proximity to the shore like geese. The Whooper Swan is one of Europe's larger species of wildfowl with a wingspan of 205–275 cm (81–108 in).

In Fife, the greatest number of Whooper Swans are usually to be seen in November and December in the vicinity of the grasslands (Elkins et al., 2003).

Figure 7.7 An East Coast White-fronted Goose (*Anser albifrons*). Note the flesh coloured bill, which is indicative of the east coast population (see text - Photo Professor Stephen Buckland).

Figure 7.8 Whooper Swans (*Cygnus cygnus*) in winter grazing by the Eden Estuary (Photo author).

Tentsmuir's thriving birds

Figure 7.9 Whooper Swans (*Cygnus cygnus*) in summer in Northeast Iceland with six goslings. This group will probably return to Scotland for the winter. (Photo Dr. H. Körner).

Figure 7.10 Bar-tailed Godwit (*Limosa lapponica*) (Photo Professor Stephen Buckland).

Occasionally, they may also be seen on Morton Lochs. Young Whooper Swans, such as those that migrate to Tentsmuir in the Autumn have mostly hatched in Iceland in July (Figure 7.9). Such is the gain in weight of young swans that at an early age they can show signs of what might be described as avian childhood obesity (Figure 7.9). With their rapid growth, the young are less agile and are particularly in need of rapid access to water to avoid falling prey to foxes.

The Whooper Swan is the European counterpart of the largest swan in the world, the North American Trumpeter Swan (*Cygnus buccinator*) to which it is closely related. When migrating south from Iceland, Whooper Swans are able to rest their heavy bodies by landing on water. Nevertheless, it has been found from observing ringed birds that some can make the journey from Iceland to Scotland without resting.

Whooper Swans are among the heaviest of all flying birds with females weighing up to 9.8 kg (21 lbs). It is therfore a source of wonder, that under favourable circumstances, these large birds can accomplish the flight from Iceland to Scotland in 13 hours.

Usually, swans fly very low, with frequent, and sometimes prolonged stops on water. However, satellite monitoring of Whooper Swans migrating between Iceland and Britain or Ireland has recorded Whooper Swans flying at an altitutude of 1856m (6089ft) above sea level (Pennycuick et al., 1999).

Shore birds

Godwits

Both the Black-tailed and Bar-tailed Godwits (*Limosa limosa* and *L. lapponica*) are regular winter visitors and passage migrants on the mud flats of both the Tay and Eden estuaries. The Black-tailed Godwit is most easily recognised by its almost straight bill (Figure 7.11), which is in contrast to that of the Bar-tailed Godwit

Common Scoter (Melanitta nigra)

Among the wintering sea ducks at Tentsmuir the Common Scoter is the only UK duck in which the drake is completely black except for a yellow patch on its beak. (Figure 7.12). The highest numbers are seen in winter, although recently they have been noted more often on autumn passage, suggesting that some of these birds have moved elsewhere to winter. Nevertheless, the most recent counts indicate a return to large wintering flocks of 3000-4000 birds in St Andrews Bay (Elkins et al., 2016a).

Common Scoters are another example of diving ducks feeding in St Andrews Bay mainly on bivalve molluscs such as mussels. These they swallow whole, subsequently crushing their shells in their gizzards. The Common Scoter has been placed on the RSPB conservation *Red List*, as in recent years there has been a greater than fifty percent decline in the already very small UK breeding population, despite the increase in the large number of wintering birds.

Figure 7.11 Black-tailed Godwit (*Limosa limosa*) (Photo Professor Stephen Buckland).

which has a slightly upturned bill. As a result Bar-tailed Godwits (Figure 7.10) favour the sand flats for feeding, while Black-tailed Godwits tend to prefer mud flats, where with their straight, long beaks, they can search for polychaete worms and especially for the Common Ragworm polychaete species *Hediste diversicolor*. For Bar-tailed Godwits, ragworms contribute the greatest part of their diet in terms of biomass, followed by lugworms (*Arenicola marina*), bivalves, snails and crustaceans. However, the latter account for less than 1% of the total diet (Bocher et al., 2013).

In Western Europe, the Black-tailed Godwit is one of the few wintering shore birds with increasing populations, probably due to the expansion of the Icelandic breeding population (Bocher et al., 2013). Tentsmuir is a relatively northern over-wintering region for Black-tailed Godwits, compared with other sections of the British coastline. It is therefore noteworthy that the Eden Estuary has recorded some of the highest numbers of this species in Scotland.

Figure 7.12 Common Scoter (*Melanitta nigra*) at River Eden Mouth (Photo Professor Stephen Buckland).

Sawbills

Among the maritime diving ducks there is one further group commonly seen in St Andrews Bay and adjacent estuaries, namely the *Sawbills*. These are slender diving ducks, where the capture of fish is facilitated by their having rearward-pointing serrations in their long bills. Regular Sawbill visitors to the Tentsmuir region are the Red-breasted Merganser (*Mergus serrator*) and the Goosander (*Mergus merganser*). Both these Sawbills (Figures 7.13 a-b) are also occasional visitors to inland freshwater coastal sites such as at Morton Lochs, as well as on the Eden Estuary, and the Firth of Tay. The greatest numbers of sawbills to be seen at Tentsmuir are the Goosanders which flock ashore at Lucky Scalp during their moulting period (Elkins et al., 2016a).

Lucky Scalp is one of a number of small islets at the north end of Tentsmuir, which from east to west – include: Larick Scalp, White Scalp, and Green Scalp.

The formation of such off-shore islets or mounds are usually due to the growth of mussel beds. It appears however, that those at Tentsmuir owe their existence to the dumping in the past of ballast taken on board up river at Perth by sailing ships after they had had delivered their cargoes. Lucky Scalp is the only islet of the group that remains above water at high tide.

The Red-breasted Merganser, which winters on coastal sites. can also be seen regularly on Morton Lochs and the Eden Estuary. It is rare to see this bird in inland sites.

The UK Sawbill populations have decreased by 31% over the past decade. This represents a severe decline from the mid-1990s when more than 1000 birds used to be observed on the Eden Estuary. Nevertheless, small parties of birds are still being reported frequenting St Andrews Bay (Elkins et al., 2016).

Figure 7. 13 (a) Red-breasted Merganser (*Mergus serrator*) (b) Goosander (*Mergus merganser*) (Photos Professor Stephen Buckland).

Figure 7.14 Male and female Shelduck (*Tadorna tadorna*) off shore at Tentsmuir (Photo Lorne Gill).

Shelduck (Figure 7.14) used to nest in the rabbit burrows at Tentsmuir Point, but since access to the sea for young birds was blocked by fencing erected by the Forestry Commission in the 1920s Shelduck have no longer nested on the dunes and moorlands at Tentsmuir (see chapter 6). Nevertheless, Shelduck continued to breed in small numbers at Earlshall in the 1960s (Summers – pers.comm.). Despite the absence of nesting birds the Eden Estuary is still one of the favoured winter sites for Shelduck in Scotland with up to 150 birds being counted on the River Eden in 2007 where Shelduck still nest in modest numbers (Elkins et al., 2016).

Birds of fresh waters

The freshwater ducks most commonly found in the fresh waters of the Tentsmuir region and adjacent areas, especially now that Morton Lochs have been restored, include Mallard, Teal, Wigeon, and Pintail. They fall into two distinct types, namely those that are surface feeders, referred to commonly as *dabbling ducks* and those that dive in search of food. Dabbling ducks have touch receptors on the inner rim of the upper bill which make it a very refined tactile organ. This, together with sensory structures on the lateral surfaces of the tongue, provide a filtering mechanism both for aquatic and terrestrial feeding (Skieresz-Szewczyk and Jackowiak, 2016).

Teal (Anas crecca)

Teal are the smallest of the dabbling ducks. They appear to owe the name *Teal* from an Old English word cognate with Middle Dutch *teling*, and Middle Low German *telink* which describes the colour of the blue-green colour around the eyes of this species (Figure 7.15).

Teal are widely distributed in Scotland as a breeding species with a preference for northern moors and mires. In the past, Teal were one of the most plentiful species of duck on the lower Tay, both as a breeding bird and as a winter migrant (Figure 7.15-16).

Figure 7.15 Head of Teal showing the dark-greenish-blue colouration around the eye that gave rise to the name of this particular colour.

Figure 7.16 Teal (*Anas crecca*) demonstrating the almost vertical take-off capacity shown by many dabbling ducks when disturbed or on the move (Photos 7.15-16 Professor Stephen Buckland).

The Victorian artist and natural historian, John Guille (1861-1935) considered *'Tentsmuir to be the foremost Teal resort in the country'* but also noted in 1876 that *'attempts to drain the Muir had greatly reduced the population of Teal'* (Millais, 1902). This drainage began with the construction of the new railway line from Cupar to Ferryport on Craig (now called Tayport see Chapter 6).

Gadwall (Anas strepera)

Gadwall (Figure 7.17) are another species of surface-feeding duck that can be seen in small numbers on the Eden Estuary and also from time to time at Morton Lochs, mainly as an autumn bird of passage. However, they have not always been part of the Scottish avifauna.

The presence of Gadwall in Scotland was first noted by Pennant in 1769 (Penant, 1771). However it was only in 1918 that they were first recorded as nesting in Scotland (Baxter and Rintoul, 1953b). In more recent times they have become evident in notable numbers as visiting autumn birds in the Tay region. When seen,

Figure 7.17 Pair of Gadwall (*Anas strepera*) on Morton Lochs. (Photo Professor Stephen Buckland).

Gadwall often occur in pairs (Figure 7.17). Gadwall remain a scarce breeding species in Fife (Elkins et al., 2016a). Nevertheless they have been observed nesting at Morton Lochs (Smout, 1986). In common with other herbivorous ducks, Gadwall do better in ponds without fish such as Carp, which compete for the vegetation.

Wigeon (Anas penelope)

Wigeon have long been known as winter visitors in Scotland. As nesting birds, however, they have only been present for a little more than 100 years. Nesting was first recorded in Sutherland (MacGillivray, 1837-1852). They then spread south through central Scotland and became plentiful resident breeding birds as well as one of our most common winter visitors (Baxter and Rintoul, 1953a).

In the early 1900s vast flocks of Wigeon began to winter in the Eden Estuary (Berry, 1939), which continues to be a well-visited site in winter (Figure 7.18). They are now generally scarce in the breeding season. Nevertheless, recent decades have seen indications of an increase in Fife, even although there has been a decline in Wigeon in Scotland as a whole (Elkins et al., 2016a).

Tufted Duck (Aythya fuligula)

The Tufted Duck is a diving duck (Figure 7.19) that favours inland fresh water lochs for breeding but moves into the Tentsmuir area at Morton Lochs in large numbers in winter. Tufted Duck have long been noted as being present at the mouths of the rivers Tay and Eden.

Figure 7.18 Wigeon (*Ansas penelope*) congregating in fresh water (Photo author).

Figure 7.19 Tufted Ducks (*Aythya fuligula*) at Morton Lochs. First discovered nesting in Scotland in 1872. Tufted ducks are an example of a duck species which benefits from living in ponds and waters that are not inhabited by fish to such an extent that there are few invertebrates left for the ducks to eat (Photo Professor Stephen Buckland).

This is another example of a duck that came to Scotland first as a winter visitor and over the past century and a half has settled and become a regular breeding species. In Fife, Loch Leven was the original main centre for Tufted Duck. However, they have now spread to many other smaller lochs including those at Morton. (Figure 7.19). The size of the wintering population in Fife has changed little in recent years.

Birds of the forest

The Tentsmuir woodlands are both extensive and varied and provide a range of tall forest trees and scrub which afford a range of habitats for a wide variety of birds.

Song birds are common throughout Fife, as well as occurring at Tentsmuir. However, it is the raptors that inhabit these coastal woodlands that are currently making Tentsmuir of growing interest both to the public in general and ornithologists in particular.

Raptors

Table 7.3 lists the Raptors (*Latin plunderer*) that have been observed in recent years at Tentsmuir and its adjacent estuaries. The most notable addition to the list in recent years has been the arrival at Tentsmuir of White-tailed Eagles which began to nest there in 2013. An extensive programme for the re-introduction

Figure 7.20 White-tailed Eagle (*Haliaeetus albicilla*) in flight over Tentsmuir (Photo John Cummins).

Table 7.3 Notable raptors recorded either as breeding birds or winter visitors in the past 5 years at Tentsmuir, Morton Lochs, and the neighbouring regions of the Tay and Eden Estuaries.

White-tailed Eagle	Peregrine Falcon
Osprey	Kestrel
Common Buzzard	Sparrow Hawk
Goshawk	Merlin

of the White-tailed Eagle to eastern Scotland in recent years has been so successful that they are now regularly seen in the Tentsmuir region.

Ospreys are also now frequently observed, but so far have not been reported as nesting at Tentsmuir, despite being given encouragement by partially preparing suitable nesting platforms. However, increasing use of the area by the Ospreys for fishing is raising the hope that they may, before long become a regular breeding species.

To look at a White-tailed Eagle (*Haliaeetus albicilla* - (Figures 7.20-21), also called an *erne* or *ern* (related to Old Norse **örn** eagle, Greek *ornis* bird) is to view a descendant of one of the most ancient groups of birds still alive today. The origin of these Sea Eagles probably took place in the general area of the Bay of Bengal where they were certainly present in the middle Miocene, 12-16 million years ago. Following prolonged persecution, White-tailed Eagles became extinct in Britain by 1918. In 1968 an early, but unsuccessful attempt was made for reintroduction. Further introductions by the Nature Conservancy were initiated from 1975 onwards. Breeding success in the Scottish population has improved over time as the average age and experience of individuals in the eagle population has increased. However, survival of released birds is lower than that of wild-bred birds, especially during the first 3 years of life (Evans et al., 2009).

Figure 7.21 White-tailed Eagles tagged as chicks at Tentsmuir in 2015 (John Cummins).

In 2013 a pair of 6 year-old White-tailed Eagles first nested successfully at Tentsmuir raising a single chick. They were the first pair of Sea Eagles to have successfully bred in the East of Scotland for over 200 years! The pair are named and tagged as Turquoise 1 and Turquoise Z (see Figure 7.21).

Since 2013 White-tailed Eagles have continued to nest regularly at Tentsmuir. The first young bird fledged there on the 15th of August 2013 and was soon seen learning to hunt with his parents. He was also fitted with a satellite tracker so his every move could followed closely by RSPB staff (see Figure 7.21).

This breeding success of the Sea Eagles at Tentsmuir continues. In 2015 to 2017 two chicks were hatched each year. In 2015 and 2016 only one chick survived. The others disappeared over grouse moors on the eastern Highlands, possibly shot illegally. In Scotland as a whole, there are now 100 nesting pairs of White-tailed Eagles (RSPB bird reports).

When feeding, White-tailed Eagles tend to perch at a suitable vantage point and watch for what ever comes into view and can be easily caught. They also have a preference for scavenging carrion rather than hunting living prey such as rabbits and fish. In this respect, in comparison with the agile Osprey, the White tailed Eagle might be described as a somewhat lazy great bird.

Osprey (Pandion haliaetus)

At the end of the 18th century the Osprey (*Pandion haliaetus*), also called Fish-Hawk or Fish Eagle (Figures 7.39-40) was common in Scotland. At that time it was particularly widespread in Inverness-shire. However by the mid 19th century the Osprey was in retreat as a nesting bird and, according to Harvie-Brown (1906), was almost driven out of Scotland as a breeding bird. Nevertheless, a few birds were recorded in the early years of the 20th century. This included a pair nesting at Loch Loyne in 1910 and possibly also a later visiting bird.

It was therefore a cause of great ornithological rejoicing in 1954 when a breeding pair of Ospreys returned to Scotland under their own volition and were reported as nesting at Loch Garten. There may have been an earlier unreported nesting the previous year. Ospreys are entirely fish-eating birds and have a specially adapted reversible outer toe which aids the

Figure 7.22 Ospreys on nest at the Loch of the Lowes where they have been nesting since 1969. Hopefully a scene such as this may become a future sight at Tentsmuir (Photo Professor Stephen Buckland).

gripping of fish. The Osprey and owls are the only raptors whose outer toe is reversible.

In diet and in hunting technique, the Ospreys differ from the White-tailed Eagles in that they are agile aerial hunters and feed exclusively on fish. The Osprey are aided in their fishing in that they have eyesight that can detect fish near the water surface even from a considerable height.

Osprey distribution

According to the Scottish Wildlife Trust there are now over 400 Osprey pairs nesting in Scotland. The number of nesting pairs in North East Europe and the Mediterranean areas has decreased during the last two decades and only the Scottish population has increased.

For a number of years Ospreys have regularly been seen hunting for fish in the Tay and especially in the Eden Estuary where flounders are the main prey. However, despite regular feeding visits every year as yet no pair has been reported as having nested at Tentsmuir. Nevertheless, there are signs that they may, in the near future, choose some part of the wooded area for bulding a nest.

At present, the main threats to Osprey well-being include pollution of the aquatic ecosystems with pesticides, habitat destruction, and losses during migration and wintering (Cieslak, 1980).

Since the Osprey started to nest in Scotland, a number of the young birds have been fitted with radio-tracking devices which have revealed that these winter migrants travel extensively, reaching West Africa (Taylor, 2010).

Common Buzzard (Buteo buteo)

In recent years the Buzzard (Figure 7.23) has become more frequent in Fife. This is especially the case in eastern Fife. They are now widespread throughout Tentsmuir. Although common in the 19th century, like Golden Eagles they also suffered from persecution. The advent of the spread of myxomatosis in the 1950s-60s and the consequent decimation of the rabbit population brought about a marked reduction in their numbers. Nevertheless, by the end of the

Figure 7.23 Buzzard (*Buteo buteo*) Photo Professor Stephen Buckland).

millennium the Buzzard population was showing a marked increase.

The Buzzards have now recovered to such an extent that their Fife population, which is largely resident, is estimated to be around 600 pairs, which represents more than a doubling of their population in the last two decades (Elkins et al., 2016a).

Buzzards like to nest at the edge of woodlands from where they can view open country in search of prey. Tentsmuir with its elongated and varied woodlands therefore provides a suitable environment for these large birds of prey.

Goshawk (Accipiter gentilis)

Although the Northern Goshawk is a widely spread species throughout North America and northern Eurasia it is a relatively new arrival in Scotland. The name in English is however ancient and can be traced back to before 1300 AD as Goshauk which comes from Old French göshafoc (about 1000 AD) meaning Goose-Hawk (Lockwood, 1993).

The Goshawk has been known to breed in Fife since 1998 and the population is thought to be between 6-12 pairs which are probably present all year round. Due to their secretive nature, the Goshawk population in Fife is probably under-recorded (Elkins et al., 2016a). In Europe it is from the coldest areas such as Northern Scandinavia, where Goshawks generally breed and from which they migrate south in winter.

The Goshawk was a favourite bird for falconers in medieval Britain, just as it is today. However, the persecution of raptors in general caused their numbers in the wild to diminish. Their recent spread northwards was however scattered and probably began with birds released by falconers in the 20th century (Petty, 1996). The recolonisation of areas from which they had vanished is thought to be at least partly due to the descendants of escapees from domestic falconry. Nevertheless, the Goshawk remains a rare bird in Scotland.

The Goshawk is by nature secretive and sightings are usually away from nesting sites. It has recently been recorded in the Tentsmuir region where it has been seen visiting Morton Lochs. The birds have a roving tendency which is a typical characteristic for this species (Elkins et al., 2016a). Being secretive however does not mean that Goshawks are timid in their hunting habits as they are capable of preying even on Kestrels and Tawny Owls (Petty, 2002).

The Goshawk is the largest member of the genus *Accipiter*. Males have a length of 49-56 cm while for females this varies between 90 and 105 cm. Similarly the wingspan of 90-105 cm for males and 108-120 cm for females is another clear example of *reversed sexual dimorphism*.

Goshawks can be found in both deciduous and coniferous forests. They seem to thrive in areas with mature, old-growth woods and are typically found where human activity is relatively low. The tree cover of Tentsmuir has probably now reached a state of maturity that makes it an attractive habitat for the Goshawk. The lack of thinning, and the partial opening of the canopy in places due to extensive windthrow, may therefore be a factor in attracting the Goshawk as it has great agility in flying and can manoeuvre skilfully between branches in a forest.

Morton Lochs restored

A recent positive action to restore the duck fauna at Tentsmuir has been the the renovation of Morton Lochs (Figures 7.24). These fresh water lochs were artificially constructed in 1906 (see Chapter 6). Together with the waters of the Eden and the Tay they now provide valuable protected territory, with a wide range of habitats sufficiently diverse to provide suitable habitats for a number of fresh water inhabiting species of ducks and other wildfowl.

At Morton Lochs the recent control of the extensive beds of Common Reeds (*Phragmites australis*) has increased the extent of open water, greatly benefitting a wide variety of aquatic birds including Kingfishers and Grey Herons and numerous species of ducks. Teal numbers have also risen as a result of the restoration, from 150 in 1998 to an estimated population of 800 during the winter 2004. Since 2004, numbers have been dropping, probably due to the fluctuating water levels. Recently however, there has now been a slight increase in wintering birds.

Morton Lochs were stocked with Carp in the early 1900s. This appears to have unwittingly also introduced Swan Mussels (*Anadonta cygnea*) which along with the fish have attracted many species of ducks.

Many other species of birds, apart from wildfowl, have been recorded frequenting these shallow lochs and the remaining roughlands of Tentsmuir. These include Redshank, Dunlin, Golden Plover, Snipe, Curlew and Lapwing (Smout, 1996).

Figure 7.24 The Morton South Loch which has recently been restored from being completely over-grown with vegetation. It is in this view that the nesting Kingfishers are being regularly observed (Photo author).

Kingfisher (Alcedo atthis)

Until recently, the presence of Kingfishers at Tentsnuir has consisted only of observations of single birds from time to time at the end of winter (Elkins et al., 2016b). The Kingfisher is very susceptable to cold weather and for this it is known in Germany as the *Eisvogel* (Ice-bird) which reflects the fact that migrants move south to Germany in response to freezing conditions to the north.

Few British Kingfishers ever move more than 250 km, though freezing weather will prompt them to move to the coast. It has been reported that severe winters can lead to as many as 90% of Britain's Kingfishers perishing. (British Trust for Ornithology, 2015). The recent arrival of several Kingfishers, all at one time at

Figure 7.25 Oblique aerial view of Morton Lochs from the north after the removal of the invasive beds of Common Reeds (*Phragmites australis*) (Photo SNH).

Figure 7.26 Kingfishers mating at Morton Lochs (a) intent, (b) event. (Photos Chris Reekie).

Morton Lochs, as noted in 2015-2016 was therefore a significant ornithological event which may reflect current climatic warming trends, and particularly the mild winters that now exist in Fife. Recent estimates had suggested that the Fife population is in the region of 30 birds (Elkins et al., 2016b), but this may now be increasing.

Kingfishers are elusive birds when nesting. They are riparian breeders, but Fife however is limited in suitable rivers for their nesting. Nevertheless it has been thought for some time that they have been nesting in the river Eden between Guardbridge and Dairsie (Elkins et al., 2003). Locating nests at Morton Lochs as a sign of breeding is extremely difficult. However, as the birds have been seen mating recently at the edge of the Morton Lochs this suggests that they are now breeding there (Figures 7.26 a-b).

The Common Kingfisher hunts from a perch 1–2 m (3.3–6.6 ft) above the water, on a branch, post or riverbank, with bill pointing down as it searches for prey. It bobs its head when food is detected to gauge the distance. It then plunges steeply down to seize its prey usually no deeper than 25 cm (9.8 in) below the surface. The wings are opened under water and the open eyes are protected by a transparent third eyelid. The bird rises, beak-first to the surface and flies back to its perch. At the perch the fish is adjusted until it is held near its tail and beaten against the perch several times. Once dead, the fish is positioned lengthways and swallowed head-first. A few times each day, a small greyish pellet of fish bones and other indigestible remains is regurgitated.

The food is mainly fish up to 12.5 cm (4.9 in) long, but the average size is 2.3 cm (0.91 in). Minnows, Sticklebacks, small Roach and Trout are typical prey. About 60% of food items are fish, but Kingfishers also catch aquatic insects such as dragonfly larvae and water beetles, and, in winter, crustaceans including freshwater shrimps.

Many young Kingfishers die within days of fledging as their first dives can leave them waterlogged and liable to drown. Because of the high mortality of young birds, Kingfishers usually have two or three broods a year, with as many as 10 in a brood (Woodall, 2001).

Tentsmuir's ornithological future

With the recent closure of the RAF base at Leuchars after a century-long presence it is possible that the use of the Tentsmuir shore may increase as a migration route and that the numbers of over-wintering birds in the area may also grow.

The high record count in April, 2008 of 450 Black-tailed Godwits on the Eden could however be due to Iceland-bound birds in passage from further south. Tentsmuir's relative proximity to Iceland in particular, and the Arctic in general, compared with more southerly wintering sites could be advantagous for the return journey northwards for over-wintering birds in spring time and account for their regular presence at Tentsmuir in winter. In the 1960s there were also records of suspected nesting of Black-tailed Godwits at Earlshall (Elkins et al., 2016a).

Such observations serve to emphasize the importance of Tentsmuir for migrating birds in general. There are few areas where within a non-built-up environment a range of suitable habitats from coastal sand-dunes to forest provide a range habitats which lie so fortuitously on major bird migration routes. The fact that over 192 species of birds have been recorded as visiting or breeding at Tentsmuir further emphasizes the importance of preserving this ecologically unique location for so many wild birds.

Chapter Eight

Tentsmuir's declining birds

Bird losses from Tentsmuir

Active conservation meausures to conserve bird populations have long been a feature of Tentsmuir's management even before the establishment there of a National Nature Reserve over 60 years ago. The management of Tentsmuir a century ago, included maintaining the duck and goose populations. To this end there was active control of many of the predators that would have had a negative effect on game birds. In addition care was also taken already in the early 1900s by Wm Berry (4th of Tayfield) to provide a range of plant species that would provide suitable forage for small birds such as Snow Buntings which are still a feature in the dune vegetation in winter (see Chapter 6).

Despite careful management from local landowners the proximitiy of Tentsmuir to centres of human population has made the bird populations vulnerable to the activities of poachers. The nadir in the mass-shooting of ducks was reached with the invention of the *punt-gun* (Figure 8.1).

This was a gun that was too big to hold and was therefore mounted on a punt. Punt-guns regularly had a bore that could exceed 50 mm and fire over 1lb (0.45 kg) of lead shot which could be capable of killing over 50 birds at one time.

The punt guns were at their most efficacious when several operated together (Newton, 2013). Teal were among the ducks that particularly suffered. Nevertheless, it was noted that many birds survived '*notwithstanding the persecution they receive from* the activities of the *punt-shooters and shore-hoppers of Dundee and St Andrews*' (Harvie-Brown, 1906). It would appear he was referring to the use of punt guns.

Given this past history of the mass shooting of ducks in places such as Tentsmuir, it is a marvel that there is still the wealth of duck species that can be seen there today. Teal still remain plentiful in flocks of several

Figure 8.1 Punt gun for the mass shooting of waterfowl (Reproduced with permission from Blue-brain.com).

hundreds at Morton Lochs as well as on the Eden Estuary. However, their numbers have been noted to be currently declining year by year, and they have not bred in the estuary for many years. Similarly, although Long-tailed Ducks are regularly seen off-shore in St Andrews Bay, and sometimes also recorded on Tentsmuir Sands, there is nowhere near the larger numbers, which used to be counted on their hundreds in the past.

In common with elsewhere in Britain, the Fife winter population has decreased and is now probably no more than 200 birds, compared with 500 birds at the beginning of the century (Elkins et al., 2016a).

Despite some local increases in a few species of wildfowl there have been many losses in the populations of breeding birds that used to nest in considerable numbers in the dune systems and slacks at Tentsmuir. The species that formerly nested regularly at Tentsmuir included Eider (Figure 8.2), Shelduck, Wigeon, Shoveler, Tufted Duck and three species of Terns; Common, Arctic, and Little. All these species occurred in considerable numbers, none of which can now be found breeding on the dunes and slacks.

The most probable cause of the decline in bird numbers is human disturbance, to which can be added fluctuations in the local availability of submerged vegetation and insect food. Shelduck (as nesting birds) were once plentiful at Tentsmuir. They first suffered a decline when the Forestry Commission removed the rabbit population whose burrows had provided suitable nesting places for these ducks.

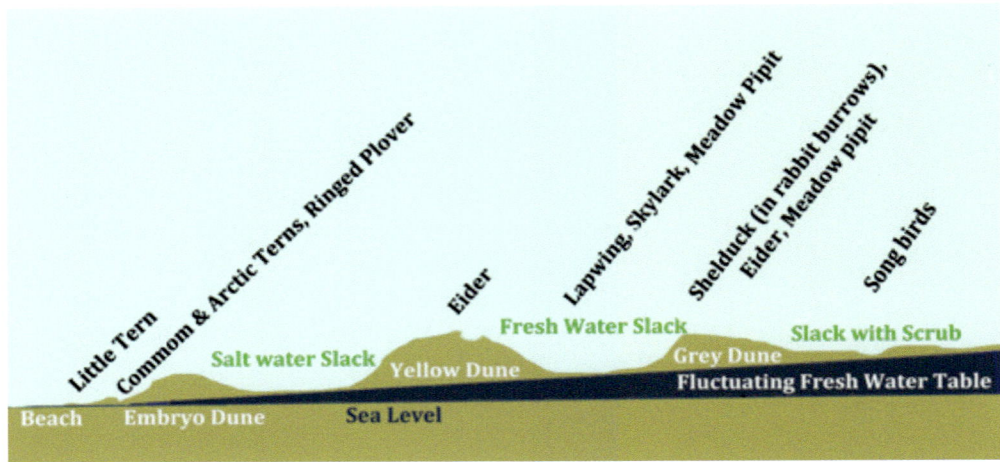

Losses of such nesting birds began to be particularly severe in the 1990s when several species, including Shelduck, disappeared almost entirely. At that time, this was described as *an ecological catastrophe* (Smout, 1996).

The fencing that was installed against the rabbits proved to be relatively unsuccessful when compared with the dramatic decimation of the rabbit population due to the myxomatosis outbreak. The presence of the rabbit fencing, was however a serious hindrance to young Eider and Shelduck from reaching the sea.

For many years Tentsmuir was also a favoured nesting place for Song Thrushes, Wood Pigeons, Yellow Hammers, Dunnock, Wrens, etc. Eider, with their large down-lined nests used to be a summer feature on the dune-heaths, drier slacks and pine plantations in the early summer. Now, with the greater human and dog presence on the moor, and predation by foxes, the nesting Eiders are no longer part of the summer scene at Tentsmuir.

Figure 8.2 Diagrammatic representation of nesting birds regularly found across the dune and slack system at Tentsmuir Point as recorded by Dr. Ron Summers in the 1960s. Songbirds included; Chaffinch, Willow Warbler, Redpoll, Long-tailed Tit, and Reed Buntings (Summers 1969).

Figure 8.3 Crèches of Eider ducklings with attendant adult females on the salt marsh of the Eden Estuary on 2nd July 1969 (Photo Dr Ron Summers).

Figure 8.2 is a diagram prepared in 1969 by Dr. Ron Summers showing the distribution of nesting birds along a transect in the region of Tentsmuir Point. At this time, the Eider, Shelduck and three species of terns were regular nesting birds on the Tentsmuir dunes and slacks. The mesh-fencing which was put in place by the Forestry Commission in the early 1920s (see Chapter 6) caused an immediate decline in the Eider duck population by preventing the young birds from reaching the sea. However, with the seaward advance of the dune system in the 1950s onwards (see Chapter 3) the Eider and Shelduck populations substantially recovered for a while and it was still possibe to see considerable numbers of Eider ducklings being gathered near the shore by their mothers in a *crèching* activity (Figure 8.3).

Although now no longer breeding at Tentsmuir, wintering Eider (Figure 8.4) can still be seen in large off-shore groups referred to as *Eider rafts*. The wintering numbers of Eider on Fife shores varies from year to year with 10,000 birds not being unusual and

Figure 8.4 Male and female Eider (*Somateria mollissima*) off shore in winter at Tentsmuir (Photo Lorne Gill).

on occasions amouting to as many as 12,250 birds (Elkins et al., 2016a). These numbers indicate that the surrounding waters at Tentsmuir can have 16% of the total over-wintering Eider UK population which is one of the greatest concentrations of Eider in Europe.

Declining wader populations

Dunlin (Calidris alpina)

There are two distinct sub-species of Dunlin, namely the subspecies *C. alpina* (Figure 8.5*)* which breeds in Russia and Northern Scandinavia, and the sub-species *arctica,* which nests in Greenland. The birds that winter in Fife belong to the Russo-Scandinavian population. (sub-species *alpina*) (while the Greenland sub-species *arctica* winters in Africa.

Although some Dunlin may be seen arriving from their breeding areas as early as July, it is September and October before the main influx arrives at Tentsmuir. In January 2009, peak numbers of 4,705 birds were recorded on the Eden, possibly at the expense of the populations on the Tay (Elkins et al., 2016). The number of Dunlin that wintered at Tentsmuir in the 1950s was regularly larger than at present. A count of 6,000 birds was made on the Eden Estuary in both November and December 1953 (Grierson, 1962).

Common Sandpiper (Actitis hypoleucos)

The Common Sandpiper (Figure 8.6) winters in Africa and is mostly seen as a passage migrant. This has long been its habit at Tentsmuir. In the 1950s the Common Sandpiper was recorded as being a migrant in small numbers on both the Tay and Eden estuaries and Morton Lochs with most records being in July and August (Grierson, 1962). It still occurs but only in small numbers. This is not surprising as its numbers decreased in Great Britain as a whole by 15 percent between 1995 and 2015. At this same time, in the 1990s, there was also a contraction in the breeding range (Harris *et al.* 2017).

Redshank (Tringa totanus)

Redshank used to be recorded as breeding in every county in Scotland on links, marshes, and rough grass and has even been found nesting up to 500 m in the Scottish Highlands. After breeding, and as soon as the young are able to fly, the Redshanks resort to the shore (Baxter and Rintoul, 1953a).

Figure 8.5 Dunlin (*Calidris alpina*) (Photo Lorne Gill).

However, Redshank have ceased to breed in Tentsmuir (Elkins et al., 2016). The number of Redshanks in Scotland increases in winter with birds that come from Iceland. In Fife the highest counts are on the Eden Estuary where flock sizes frequently exceed 1000 birds. Numbers peak in November at Tentsmuir where the size of the population has held up despite a 21% fall in winter in the UK population as a whole (Frost et al. 2017).

Curlew (Numenius arquata)

One of the most notable declines in recent years has been the fall in the number of Curlews (Figure 8.7) which have have fallen by 17 % in the last decade. This notable decline is probably due to the eradication of much of the rough pastures that used to provide suitable breeding areas in Scotland for large numbers of these birds. It is in the breeding sites on the Scottish uplands that some of the greatest environmental changes have taken place due to the extensive agricultural improvement of the pastures. This has resulted in a widespread decline in the total breeding population of Curlew (Douglas et al., 2014), with the species suffering a 59% decrease across Scotland between 1995 and 2015 (BBS, 2013). Similarly, the number of breeding birds at Tentsmuir has also suffered a decline. This has been due to the increasing popularity of Tentsmuir and its Nature Reserve reserve for human visitors and their dogs. The degree of disturbance is generally too great for many nesting birds and Curlews have now abandoned Tentsmuir completely as a breeding area.

Figure 8.6 Common Sandpiper (*Actitis hypoleucos*) (Photo Professor Stephen Buckland).

Figure 8.7 Curlew (*Numenius arquata*) (Photo author).

The breeding population in Fife as a whole is now fewer than 50 pairs as compared with 730 pairs recorded in the 2003 edition of the Fife Bird Atlas.

Oystercatcher (Haematopus ostralegus)

In the east of Scotland the breeding behaviour of Oystercatchers (Figure 8.8) differs from those in the west, where their main breeding areas have long been on shores and rocky islets. It was already noted (MacGillivray, 1837-1852) that *along the eastern part of the coasts of Scotland, few breed on the rocky headlands the greater part betaking themselves to the rivers, on the stony or sandy beaches of which they form their nest.* Oystercatchers do not nest by rivers in Fife. Instead, they use arable fields and some even nest on flat rooves. Half a century ago the number of wintering Oystercatchers counted on the Eden Estuary alone could number 5,000 birds (Grierson, 1962). Over 2000 birds are still currently counted regularly on the estuaries of the Eden and Tay with a total wintering population in Fife estimated at 6000 birds (Elkins et al., 2016).

The latest counts collected by the Wetland Bird Survey (WeBS) has revealed a continuation of the sharp drop in the number of Oystercatchers wintering in Scotland. Having steadily increased during the 1980s and 1990s, numbers of Oystercatchers wintering in Scotland have declined in recent years at a rate which is greater than the earlier increase.

In relation to breeding birds, between 1991-99 and the latest 2008-13 survey the number of breeding pairs of Oystercatchers in Fife fell from 1400 to 400 Understanding the causes of this change in the fortunes of Oystercatchers in Scotland as a whole,

may be crucial in ensuring that they remain a key feature of Scottish wetlands in the future (Elkins et al., 2016).

Figure 8.8 Oystercatcher (*Haematopus ostralegus*). (Photo Lorne Gill).

Sanderling (Calidris alba)

In contrast to the losses noted above there are however, some species in recent years which have shown an increase in their wintering populations in Scotland. Sanderling is such an example (Figure 8.9) showing a very pronounced increase from the mid-1990s. This species of wader which breeds in the High Arctic has long been a common shorebird in winter in Scotland. In the 19th century they were found both on the west coast of Scotland and the

Figure 8.9 Sanderling (*Calidris alba*) (Photo Lorne Gill).

Figure 8.10 The Grey Plover (*Pluvialis squatarola*) in winter plumage with a brown-grey colouration having lost the blacker feathers of its summer plumage (Photo Professor Stephen Buckland).

Firth of Forth. By the 20th century Sanderling had also become increasingly common on the Scottish east coast. In contrast to the Grey Plover, which likes muddy shores, Sanderling, as their name implies, prefers sandy shores. In Fife, the favourite winter locations are Largo Bay and Tentsmuir Sands. Some Siberian birds stage and moult in North West Europe before continuing to South Africa along the Eastern Atlantic Highway.

On the non-estuarine Scottish coasts some estimates put the number of Sanderling migrants possibly as high as 10,000 birds. This number however includes a large number of winter birds in the Outer Hebrides

Plovers

Grey Plover (Pluvialis squatarola)

The Grey Plover (Figure 8.10) is a regular winter visitor and bird of passage at Tentsmuir where the largest count has been of 250 birds on Tentsmuir Sands in February 2008. Large muddy estuaries are their preferred feeding grounds and they are found mainly on the estuaries of rivers, such as the Tay and the Eden and similar habitats elsewhere in eastern Scotland where they tend to feed on Ragworms (*Hediste spp.*). The total Fife population in winter can number 1000 individuals (Elkins et al., 2016).

Ringed Plover (Charadrius hiaticula)

Ringed Plover coastal breeding populations have declined in recent years in Fife mainly due to disturbance and predation although inland populations appear to remain stable. Outside the breeding season in Spring considerable counts of Arctic migrants have been made. Notably one in 2009 of 604 birds at Tentsmuir Point (Elkins et al., 2003). On the Tentsmuir Nature Reserve, Ringed Plover still breed along the dune edges and can be seen scuttling about the foreshore and dunes.

The N.E. coastal area of Holland is now proving less suitable for mollusc-feeding birds such as Oystercatchers (Schwemmer et al., 2014). Recent counts have found that the more dependent migratory birds are on the Wadden Sea, the less successful they are in maintaining their numbers. (Blew et al., 2016). The suggested explanation is that the deep dredging that has been taking place there in recent years has damaged the shellfish feeding and this has brought about the decline in mollusc-feeding birds on the Wadden (van Gils et al., 2006).

The reasons for the decline of some of the most abundant wintering waders on UK estuaries are not fully understood, but it is likely to be due to a combination of factors. Waterbird counts from across north-west Europe also show that species are shifting north-eastwards in response to milder winters. In addition, the number of breeding birds has also declined, particularly in upland areas where extensive modern pasture management techniques for cattle grazing in recent years have reduced their feeding value and nesting suitability for birds.

Wigeon

In the early 1900s vast flocks of Wigeon began to winter in the Eden Estuary (Berry, 1939), which continues to be a well-visited site in winter. They are now generally scarce in the breeding season. Nevertheless, recent decades have seen indications of an increase in Fife, even although there has been a decline in Wigeon in Scotland as a whole (Elkins et al., 2016a). The size of the wintering population in Fife has changed little in recent years and remains between 1500-2000 birds (Elkins et al., 2016).

Historical bird records from the Eden Estuary

Monthly counts were made between 1950 and 1955 on the Eden Estuary of 7 of the most common species of ducks and 6 of the most common species of waders (Grierson, 1962). When compared with more recent counts these figures show (Figure 8.13) that it has long been apparent that many of the most familiar wading birds flocking to UK estuaries in winter have declined greatly in their numbers in recent years. The British Trust for Ornithology has been monitoring wetland bird numbers on the East Atlantic Flyway since the early 1980s with their Wetland Bird Survey WeBS (see www.bto.org). These observations have noted a persistent decline over this period in the populations of Ringed Plovers and Dunlin.

By contrast, Black-tailed Godwits stand out as the one species that has shown an increasing winter presence as a whole during this period. Nevertheless, roosting flocks of Bar-tailed Godwits which regularly exceeded 1500 individuals on the Eden Estuary in the early 20th century now rarely number more than 1000 (Elkins et al., 2016).

The breeding population of Black-tailed Godwits in Fife is now fewer than 50 pairs as compared with 730 pairs recorded in the 2003 edition of the Fife Bird Atlas.

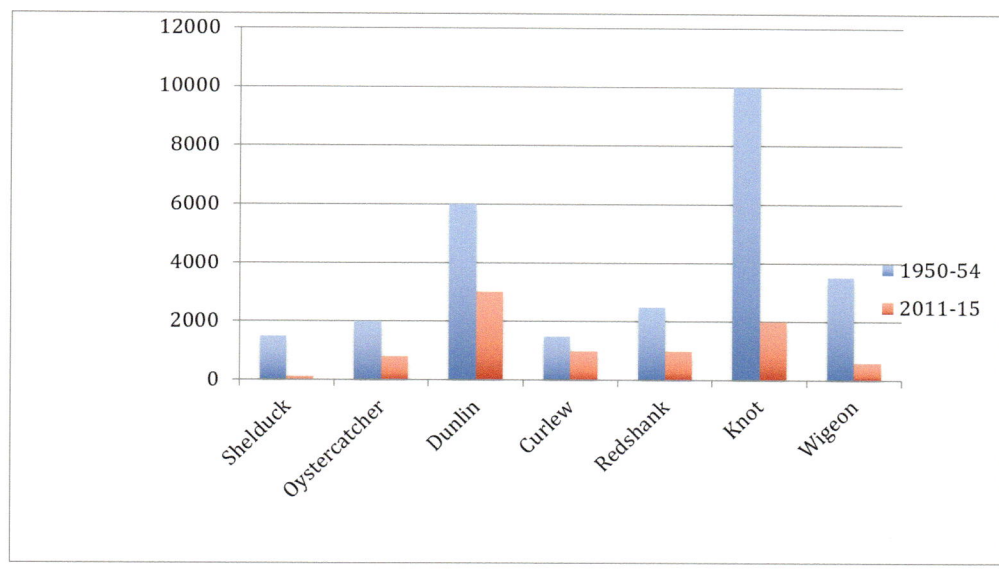

Figure 8.13 Examples of the decline over the past 60 years of over-wintering shore birds on the Eden Estuary. Bird numbers for 1950-54 are from approximate maximum monthly counts (Grierson, 1962). Figures for 2011-2015 are approximate maximum winter records (Elkins et al., 2016a).

Wildfowl and lead poisoning

The long history of discharging shot guns, and losing lead weights in fishing etc. in wetlands, rivers and lakes, has created a significant presence of lead in the aquatic environment. Large numbers of Wigeon are likely to die each winter (i.e. during the shooting season) as a direct result of lead poisoning. Wildfowl that die outside of the shooting season will be additional, as will those that die of causes exacerbated by lead poisoning.

Lead levels are significantly negatively associated with winter body condition when they rise above 44µg dL^{-1} of blood in winter. These findings indicate that sub-lethal levels of lead impact on body condition at the lower end of previously suggested clinical thresholds. Broad estimates indicate that in the UK, in the order of 50,000-100,000 wildfowl (c. 1.5-3.0% of the total wintering population) are affected by lead poisoning.

These findings serve to reaffirm the importance of reducing the contamination of lead in the environment. The larger the birds, the greater the risk of absorbing a toxic amount of lead. For migratory swans, this represents a quarter of all recorded deaths. (Newth et al, 2011, Krone et al, 2009).

Physiology of swan-song

The term 'swan-song' derives from the legend that, while Mute Swans are relatively mute during life, yet they sing beautifully and mournfully just before they die. This is not actually the case. The legend has long been known to be false, as even at the time of Ancient Rome, Pliny the Elder (AD 23-79) refuted it in his Natural History (AD 77, book 10, chapter xxxii - a work that was completed just 2 years before his death in the AD 79 eruption of Vesuvius).

'olorum morte narratur flebilis cantus, falso, ut arbitror, aliquot experimentis'
(*The account of this doleful death-song is false and without any experimental evidence*).

However, the Whooper Swan does possess a 'bugling' call, and has been noted for issuing a drawn-out series of notes as its lungs deflate upon expiry. When it was the habit to shoot anything that flew, there would have been many occasions when hunters would have encountered dying swans. The Whooper Swan and its close relatives the Trumpeter and Tundra Swans have convoluted wind pipes due to an additional tracheal loop around a deep hollow in the sternum. The German zoologist and botanist Peter Pallas (1741-1811) was the first to propose this as the explanation for the *Swan-Song* legend of the dying swan. For these large swans, their expiring breath can create these dying notes while passing through the anatomical resistance of a convoluted windpipe.

Chapter Nine

Tentsmuir's mammals, butterflies and moths

The past century has seen many physical changes at Tentsmuir as the coastline has changed and with it the nature of the vegetation cover has altered extensively. This chapter therefore explores the effects that these environmental alterations have had on some of the fauna of Tentsmuir. Just as Tentsmuir provides a habitat for many species of birds (see Chapter 7), it also provides suitable territories for other sections of the native fauna. These range from the foreshore with its Grey and Harbour Seals, to the dunes and grasslands with their butterflies and moths for which it has long been famous.

When Edward Wilson (1872-1912) visited Tentsmuir in June 1909 (see Chapter 6), the year before he left with Scott on the ill-fated Terra Nova Antarctic expedition, he made a note in his diary that the butterflies and moths he had seen were a significant feature of the day's outing (Seaver 1933). Despite the monotony of the extensive present forestry plantation there can still be found a diverse fauna of 19 different species of butterflies and over 320 species of moths, all of which exhibit distinctive habitat preferences. The forest also has its own special attractions among which there is a lively population of Red Squirrels moving through the branches and along the ground in search of food. On land, the most recent new arrivals are the introduced Beavers, which will if their numbers increase, undoubtedly have an impact, not only on Tentsmuir's wetlands but possibly also on the forest.

Seals

Extensive erosion has removed many of the sand bars where the large numbers of seals used to be readily observed. However, despite these changes in the foreshore both Harbour (Common) and Grey Seals are still present Figures 9.1-2 and 9.4), although less numerous than in the past. One of the attractions of late summer at Tentsmuir used to be the sound of the calling of the pups of Grey and Harbour Seals when, along with the adults, they hauled-out on the sandbanks.

On a still morning in late summer, the sound of howling pups coming through the mist from the off-shore banks could strike a uniquely melancholy note. The author recalls a North German visitor in August 1964 saying, when he heard these haunting howls, that at this time of the year in Germany, a seal pup is called a *Heuler*, (instead of the normal German term of *Seehundbaby*).

In the 1960s, the sand-banks were assumed to provide a secluded location free from excessive human disturbance and therefore ideal places for resting seals and howling pups. Unfortunately, such idyllic tranquillity is now only too easily and frequently disturbed. Increased numbers of irresponsible visitors, many of whom are accompanied by more than one unleashed dogs, who venture out onto the sand

Figure 9.1 Grey Seals hauled out on sandbank at Tentsmuir (Photo SNH).

banks at low tide and cause a considerable degree of disturbance of the hauled out seals.

Grey Seals

Grey Seal numbers have remained little changed in recent years. A survey carried out in 1995-1996 (Figure 9.3) estimated the total count for Scotland to be 21,602. A repeat survey carried out between 2011-2015 produced only a slightly smaller number of 18,968 (Duck and Morris 2016).

When detailed totals for separate years for the Firth of Tay and Eden Estuary are examined (Figure 9.3) it can be seen that despite some variations from year to year, there is no large overall difference in the total number of Grey Seals as recorded in the early 1990s compared with the last decade. (Figure 9.3).

The Grey Seal is territorially more mobile than the Harbour Seal. As a result, the numbers of Grey Seals hauling out can be highly variable in the summer

Figure 9.2 Grey Seal pup (SNH).

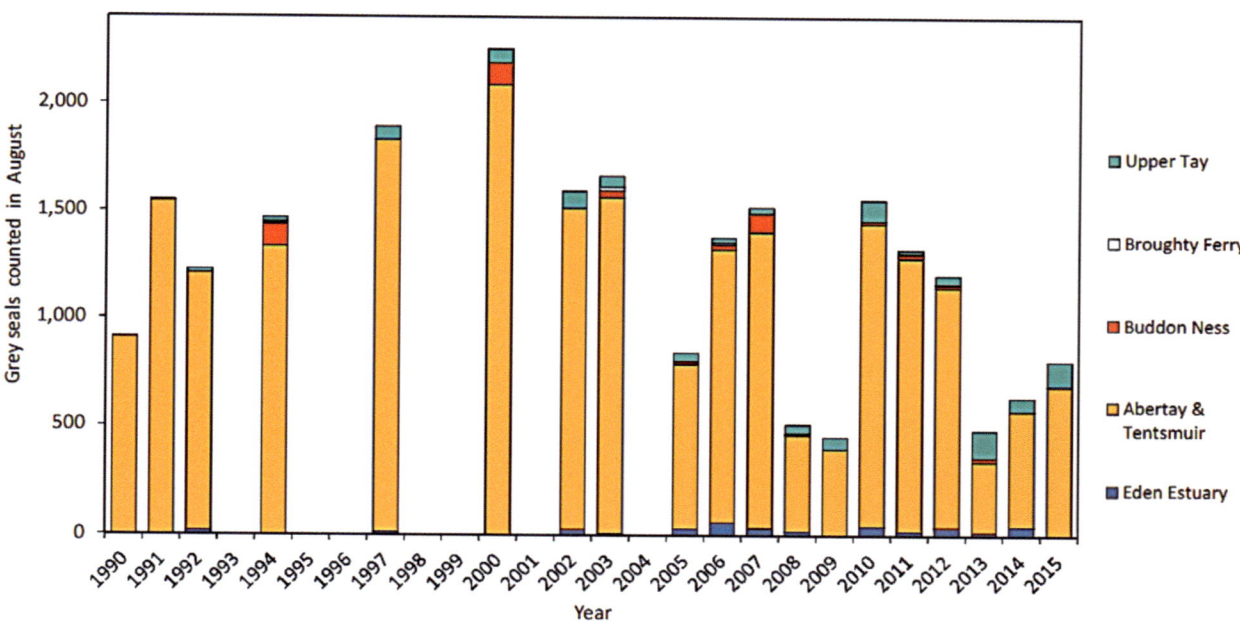

Figure 9.3 August counts of Grey Seals in the Firth of Tay & Eden Estuaries from 1990 to 2015. Mean values are given for areas surveyed more than once in aerial surveys by the Sea Mammal Research Unit (Duck and Morris 2016). Reproduced with permission from Scottish Natural Heritage.

Figure 9.4 Harbour Seal (*Phoca vitulina*) on beach at Tentsmuir (Photo Lorne Gill-SNH).

months and caution is advised when interpreting local population variations (Duck and Thompson 2008).

Harbour (Common) Seals (Phoca vitulina)

In recent years there has been a marked decline in the number of Harbour Seals. However, this cannot be blamed entirely on human disturbance.

The Sea Mammal Research Unit at St Andrews University, has been monitoring the Scottish seal numbers and has noted a steady decline of the population of Common Seals on the east coast of Scotland (Figure 9.5). The deline in the number of Harbour Seals in the Tentsuir area, from the Estuary of the Upper Tay to the Eden Estuary has been particularly notable.

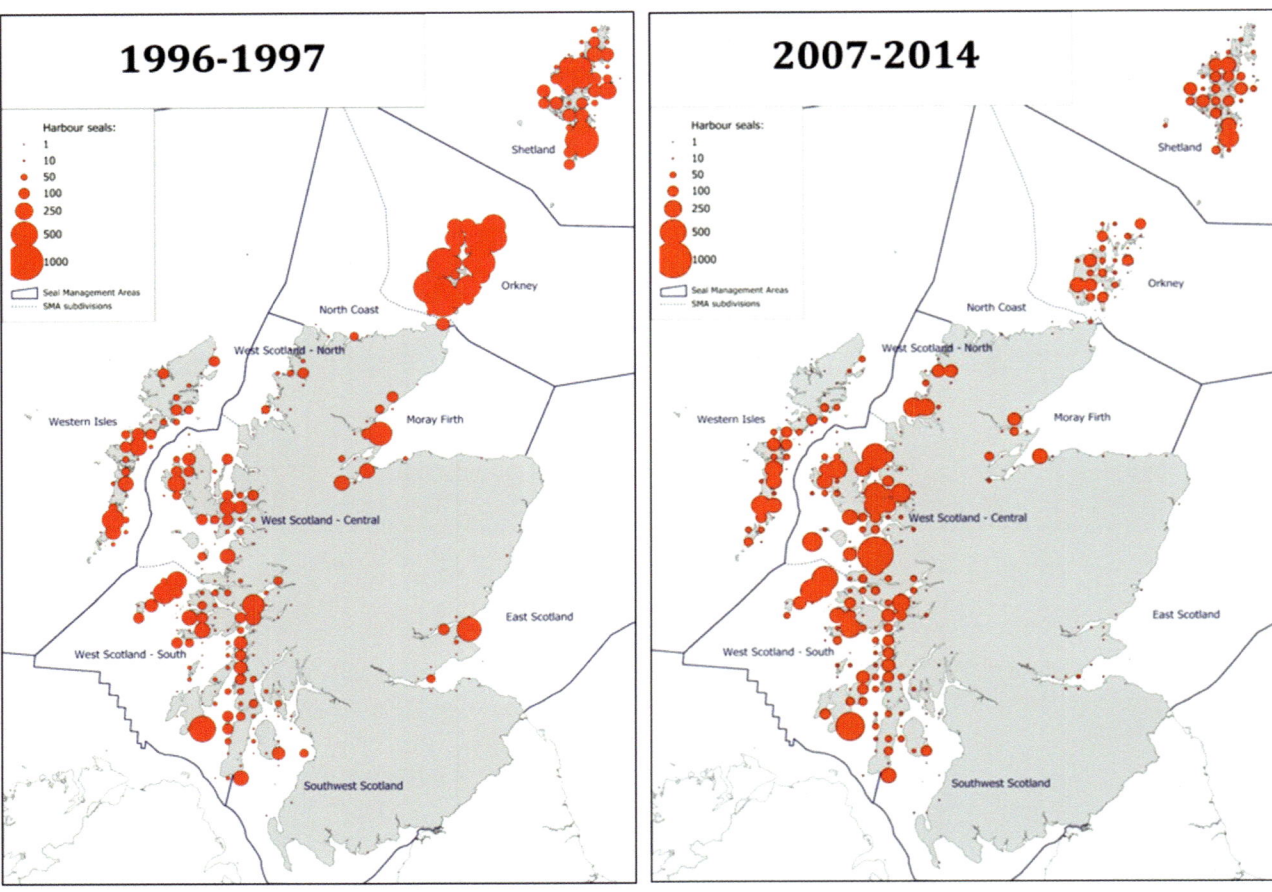

Figure 9.5 Changes in population size and distribution of Harbour Seals in Scotland of counts in 1996-1997 and 2007-2014 (Duck and Morris 2016). Reproduced with permission from Scottish Natural Heritage.

This decline contrasts with the relative stability of British Grey Seal populations as a whole.

The decline in the shore counts of Scottish Harbour Seals have also been verified by detailed and careful monitoring from the air. From these observations it appears likely (Figures 9.6-7) that the Harbour Seals in the region between the Rivers Tay and Eden may become extinct in the near future (Hanson, Thompson et al. 2017).

Where just under 400 Harbour Seals used to be counted regularly, the number is now less than 200. This general decline in Harbour Seal numbers contrasts with the relative stability of British Grey Seal populations as a whole. At Tentsmuir the decline in seal numbers on the off-shore sand banks is so severe that in 2016 only one Harbour Seal pup was observed during September and this was one of only 12 to be observed on the east coast of Scotland as a whole (Cunningham 2015-16).

Tentsmuir's mammals, butterflies and moths 145

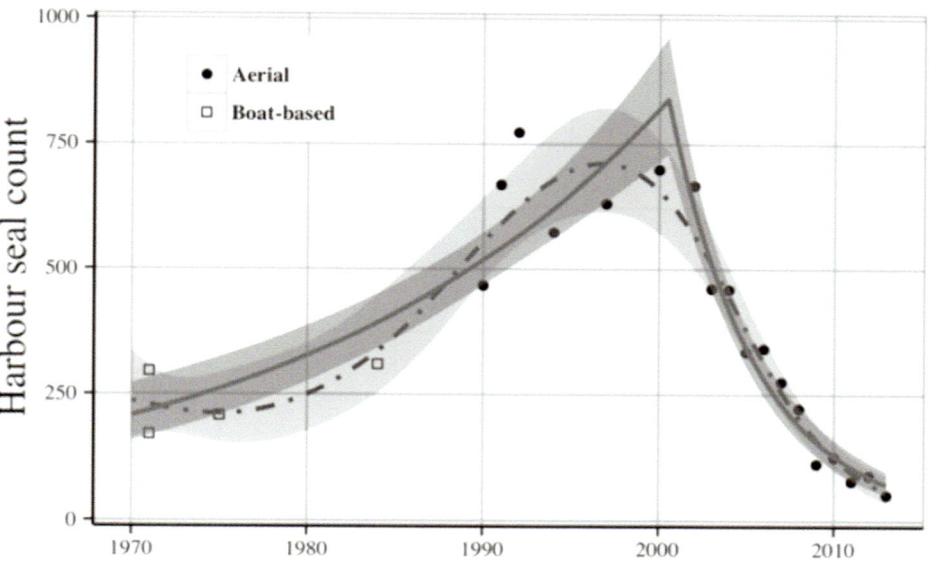

Figure 9.6 August counts of Harbour Seals in the Firth of Tay & Eden Estuary from 1990 to 2015 (Data are from aerial surveys by the Sea Mammal Research Unit. (Duck and Morris 2016) Reproduced with permission from Scottish Natural Heritage.

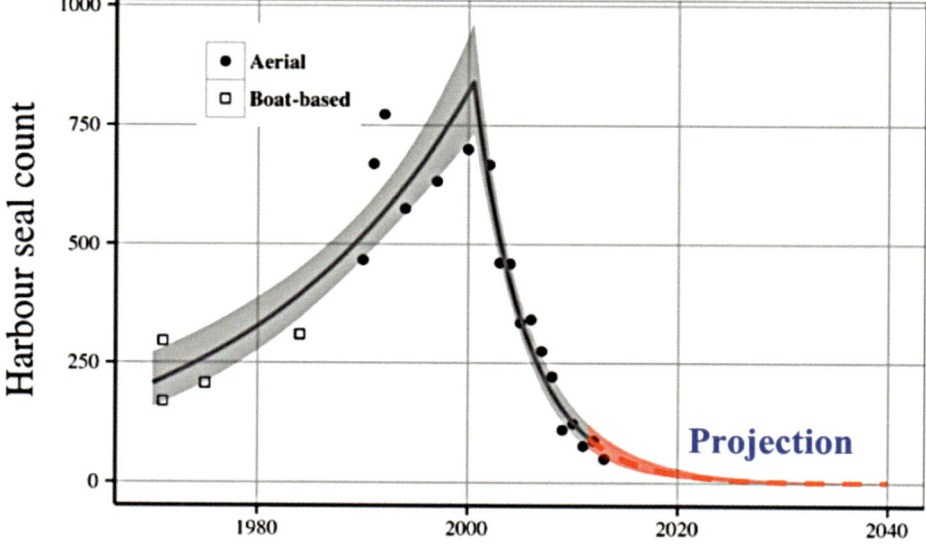

Figure 9.7 Projection from Gatty Marine Laboratory model with 95% confidence limits of the Harbour Seal counts in the Firth of Tay and Eden Estuary shown in red after 2013 (Reproduced with permission from Hanson et al, 2017).

The decline in the shore counts of Scottish Harbour Seals has been verified by detailed and careful monitoring from the air From these observations it appears likely (Figures 9.6-7) that Harbour Seals in the region between the Rivers Tay and Eden, may become extinct in the near future (Hanson, Thompson et al. 2017). It was however concluded, that pup mortality could not be the primary driver of the decline, as the survival of the young in falling populations was similar to that when it is stable (Hanson, Thompson et al. 2017).

Fresh light was shed on this problem by the finding in Canada of Harbour Seal carcasses with wounds which were described as flesh-lacerations which appeared to suggest another possible cause for Harbour Seal decline. Post-mortem examination of such wounded Canadian Harbour Seals revealed they had been alive and healthy when the injuries were sustained, and that there was no evidence of any underlying disease or disability.

It was suggested that the cause of death of these seals, with such characteristic spiral lacerations washing up on Canadian shores, over the past two decades was caused by a sudden traumatic event involving a strong rotational shearing force. The injuries were therefore thought to be consistent with the animals being drawn through the ducted propellers of marine vessels (Bexton, Thompson et al. 2012). The possibility that these helical-shaped wounds were being caused by interactions with ship propellers lead to them being described as *corkscrew injuries*. This challenged the conclusions of a previous study in Canada that suggested natural predation by Greenland Sharks (*Somniosus microcephalus*) was responsible for these injuries (Nielsen, Hedeholm et al. 2014) .

However, new evidence from observations in Scotland points towards another more likely explanation, namely Grey Seal predation. On the Isle of May, an adult male Grey Seal had been observed and recorded catching, killing, and eating five weaned Harbour Seal pups over a period of one week (Brownlow, Onoufriou et al. 2016). A further 9 carcasses found in the same area exhibited similar injuries. Post mortem examination of the lesions indicated that the wound characteristics were similar to each other. In 12 of the 14 carcasses analysed, the injuries were indistinguishable from carcasses previously attributed to supposed propeller injuries and described as *corkscrew injuries*.

It is therefore probable that most of the seal carcasses displaying spiral lacerations, at least in the UK, are caused by Grey Seal predation and that similar cases in other locations should be re-evaluated to identify whether or not Grey Seal predation is a widespread cause of mortality in Harbour Seals. This is needed as since 2010 harbour seal carcasses with characteristic spiral wounds continue to be recorded in the vicinity of the Firth of Tay and Eden estuary. This level of mortality is estimated to be unsustainable, and likely to be a major factor in the decline of the Harbour Seals (Brownlow, Onoufriou et al. 2016).

Lepidoptera

As with most wild-life counts, the different species of butterflies and moths, observed at Tentsmuir tend to show preferences for particular habitats with some species favouring the forward exposed dunes and slacks, while other choose the more inland and sheltered forest and the wetlands that occur around

Figure 9. 8 Photo of Len Fullerton (1909-1968) mounted into one of his paintings of Mute Swans. He was an outstanding naturalist and wildlife artist who was particularly fond of Tentsmuir and Morton Lochs. His daughters recalled that he cycled down to Tentsmuir from Newport on Tay every Sunday where his friend Ellis Crapper joined him. Crapper's herbarium of the Tentsmuir Flora and is now lodged in a Tentsmuir collection in the Archives of St Andrews University along with many of Len Fullerton's paintings and sketches. (See Crapper, 1939).

areas such as Morton Lochs. The present count for Tentsmuir's Lepidoptera amounts to 19 species of butterflies and 320 or more species of moths (Bryant 2017).

Butterflies

Regular and systematic recording of butterflies at Tentsmuir began soon after the establishment of the *United Kingdom Butterfly Monitoring Scheme* in 1976. This coincided with the appointment of Peter Kinnear as the National Nature Reserve manager for Tentsmuir and Morton Lochs.

In an early description of Tentsmuir's dunes, Len Fullerton (Figure 9.8) a local naturalist and artist with a long-standing interest in the area, gave a vivid description of the butterfly species that were associated with the dune valleys (Fullerton 1956). He noted in that the the fast-flying Dark Green Fritillaries

Kinnear, in common with other recorders, gave particular attention to the butterflies of the forward dunes and heathlands, which were monitored along a 4 km long transect from 1978 until he retired in 1996. What was decribed as a Provisional Atlas of the butterflies of Fife was published at the end of the last millennium. Examination of the maps in this atlas (Smout and Kinnear 1993) is more than a record of

Figures 9.9 (a-b) Dark Green Fritillary (*Agrynnis aglaja*) (Photos - Professor Stephen Buckland).

Figure 9.10 Past distribution in Fife of the Dark-green Fritillary (Smout and Kinear 1993).

(*Agrynnis aglaja*) as being particularly conspicuous and plentiful as they careered about the dune-valleys and fed on the thistles and thyme where they tended to settle (Figure 9.9 a-b).

Fullerton also noted that the Grayling, as being in the habit of canting at an angle to the sun reduced their shadow and made them difficult to locate on the dry heath on which they tended to settle.

Figure 9.11 Grayling Butterfly (*Hipparchia semele*), (Photo Professor Stephen Buckland).

Two of the most frequent species of butterflies noted on Kinnear's transect throughout this period were the Grayling (*Hipparchia semele* (Fig. 9.11), and the Small Copper (*Lycaena phaeas* Figure 9.12). These two species stand out in the 1993 distribution maps as characteristic of the dunes and heaths of the nature reserve. The Grayling is less colourful than the Small Copper but if startled, will flick its forewing upwards to reveal an upper hindwing with a distinctive eyespot. In terms of distribution, the Grayling (Figure 9.11) is restricted in Fife to the Tentsmuir area while the Small Copper (Figure 9.12) although very frequent in Tentsmuir is restricted mainly to the north of Fife (Smout and Kinnear 1993).

The Common Blue Butterfly (*Polommatus icarus*) has a very distinct distribution occuring throughout Fife on dunes and heathlands (Figures 9.14-16) where it is a particularly characteristic species of the dune and heath areas. The male of this species is conspicuously blue. The colour of the female varies from partly brown and partially blue early in the year to a more predominantly blue colour later in summer. The metallic blue colour which is seen best in the male (Figures 9.15) is not due to pigmentation, but due to diffraction of sunlight by thousands of corrugated scales which differentially absorb and reflect light on the wings of the males.

Adult Common Blue Butterflies drink nectar from flat-headed flowers. The caterpillars of this species eat wild leguminous plants such as Bird's Foot Trefoil, White Clover and Rest Harrow. The vegetation of the open grasslands provides good feeding for the young caterpillars.

Figure 9.12 Small Copper Butterfly (*Lycaena plaeas*). This image is of a female, which is slightly larger than the male and has rounded forewings. (Photo Professor Stephen Buckland).

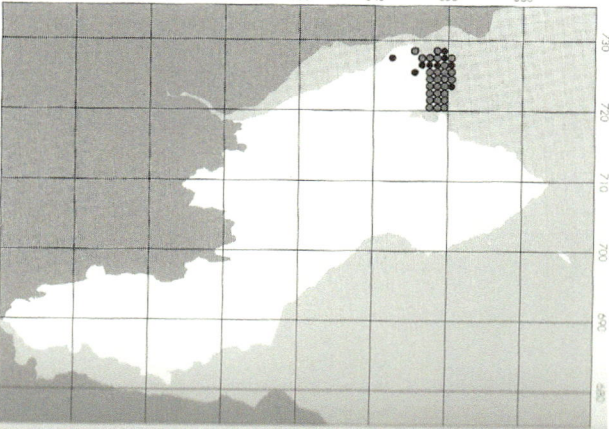

Figure 9.13 Past distribution of Small Copper in Fife (Smout and Kinnear, 1993).

the butterflies present in Fife at that time. It is also a remarkable ecological record of species habitat preferences and reveals how particular some species of butterflies can be in their habitat preferences.

Figure 9.14 Common Blue Butterflies in Fife. (a) Female early in season. (b) Female later in summer. (Photo Professor Stephen Buckland).

Figure 9.15 Male Common Blue Butterfly (*Polommatus icarus*) (Photo Professor Stephen Buckland).

The Orange Tip butterfly (Figure 9.19) although not numerous at present is included as it apppears to be a species which is responding to climatic warming by becoming more frequent in some Scottish habitats. Its preference for wet meadows and the loss of such areas due to modern agricultural practices of improved drainage and silage cutting is however making it less

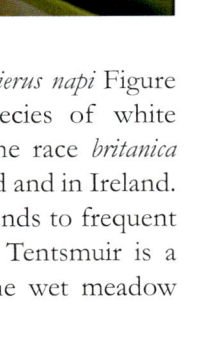
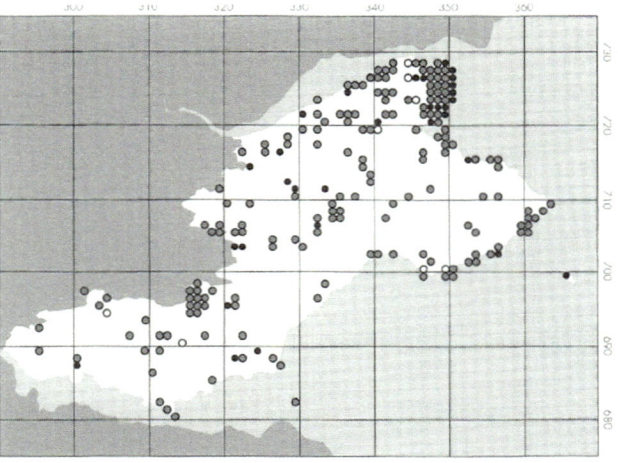

The Green Veined White Butterfly (*Pierus napi* Figure 9.17) is one of the commonest species of white butterflies in northern Britain with the race *britanica* being the form which is found Scotland and in Ireland. It occurs in a variety of habitats but tends to frequent wet meadows and woodland fringes. Tentsmuir is a particularly favourite location with the wet meadow vegetation in its dune slacks.

Figure 9.16 Past distribution in Fife of Common Blue Butterfly (Smout and Kinnear, 1993).

Figure 9.17a-b (Green Veined White Butterflies (*Pieris napi ssp.brittanica*) at Tentsmuir (a) example of female from a spring brood (b) summer brood (Photo Professor Stephen Buckland).

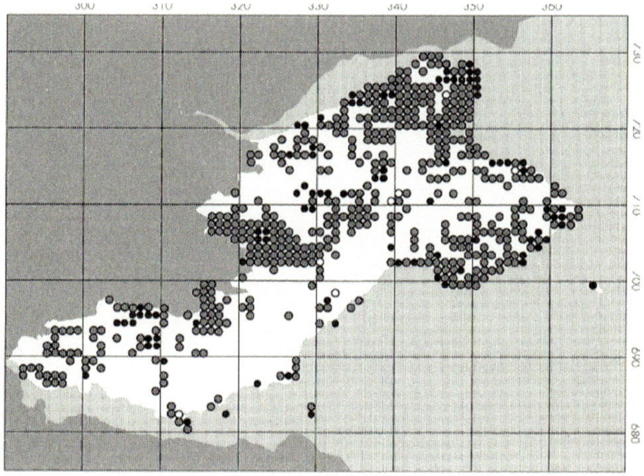

Figure 9.18 Past distribution in Fife of Green Veined White Butterfly. (Smout and Kinear 1993).

Figure 9.19 Orange Tip Butterfly (*Anthocaris cardamines*) at Tentsmuir (Photo Professor Stephen Buckland).

common in the more southerly locations. Nevertheless, the species still abounds in the wet meadows around the dune slacks and woodland glades of Tentsmuir. (Newland and Still 2010).

The Ringlet (*Aphantopus hyperantus*) (Figures 9.20-21) like the Dark Green Fritillary (*Argynnis agaja*) is another

species of butterfly which is particularly common at Tentsmuir. Like the Orange Tip (Figure 9.20), it is most commonly seen in colonies inhabiting woodland rides and lush grasslands, where it flourishes, provided the area it is not subject to prolonged droughts and overgrazing (Newland and Still 2010).

Figure 9.20 The Ringlet (*Aphantopus hyperantus*) (Photo Professor Stephen Buckland).

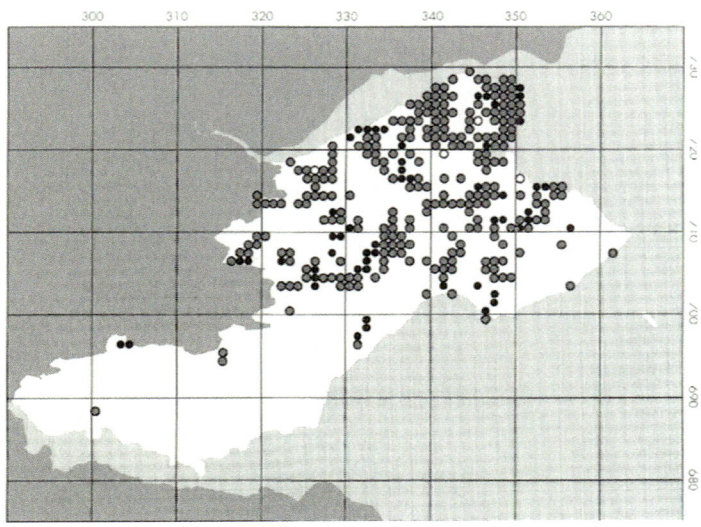

The Small Heath Butterfly (*Coenonympha pamphilis* Figures 9.22-23) is also a species that frequents the dunes, slacks and woodland clearings at Tentsmuir as well as occuring throughout Fife.

The Painted Lady Butterfly (*Vanessa cardui*) (Figure 9.24) is a long-distance migrant, which arrives in vast numbers in the British Isles each year, spreading northwards from the desert fringes of North Africa in the area of Maghreb, (variously referred to historically as Barbary, Mauretania, Numidia, and Libya). It is an example of a butterfly species that is now becoming more common

Figure 9.21 Past distribution of the Ringlet in Fife (Smout and Kinnear 1993).

Figure 9.22 The Small Heath Butterfly (*Coenonympha pamphilis*) (Photo Professor Stephen Buckland).

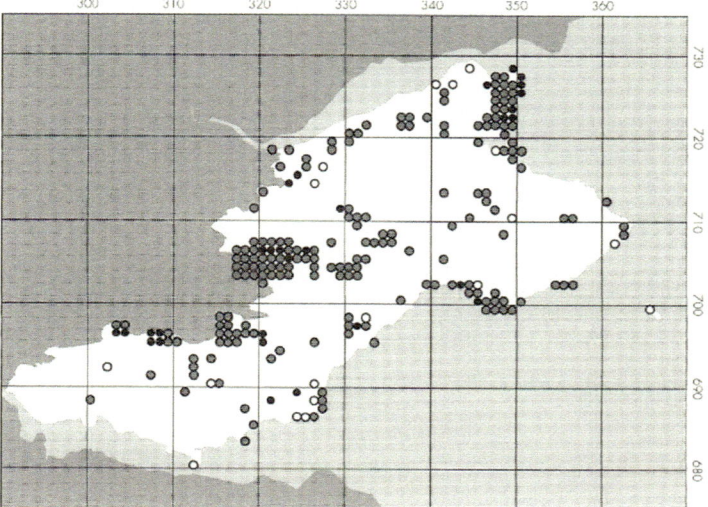

Figure 9.23 Past distribution of the Small Heath Butterfly in Fife (Smout and Kinnear 1993).

at Tentsmuir. As the northward migration takes place in spring, succesive populations breed and then progressively migrate futher and further north. Painted Ladies are also now becoming regularly recorded at Tentsmuir, probably as a result of climatic warming.

Until recently it was not known if the Painted Lady, like the closely related Red Admiral, made the return journey to North Africa at the end of the summer, or simply died in the UK. It had been generally supposed that the northern outlying populations of Painted Lady butterflies failed to return to Africa.

Investigations using a special harmonic form of radar that can detect butterlies in flight have shown that the Painted Lady Butterflies do make a return flight to Africa flying at altitudes up to 3,000 feet, at speeds reaching up 30 mph and cross the Sahara Desert.

Declining butterfly populations

While butterfly counts remain stable for many species at Tentsmuir, there are nevertheless examples of significant reductions in others (Figure 9.25). The Grayling, the Small Copper, and the Dark Green Fritillary populations at Tentsmuir have all suffered significant declines in recent years. By contrast the Common Blue and the Painted Lady populations have increased.

Moths

Tentsmuir has long been noted for its many species of moths. Taxonomically, they can be divided into the Macro-moths and Micro-moths. At present Scotland has 560 species of Macro-moths out of a total of 900 recorded for Great Britain as a whole (Bryant and Gerald 2017). Len Fullerton a local naturalist and artist who made a lifetime study of Tentsmuir (Figure 9.8) with many notes on the Butterflies of Tentsmuir (see above) also remarked on its rich Macro-moth fauna (Fullerton 1956).

In addition to moths being divided into macro and micr-species they can also be described as those that fly by day and those that fly by night. Some examples of the commoner day-flying macro-species are given below.

Figure 9.24 The Painted Lady (*Vanessa cardui*) (Photo author).

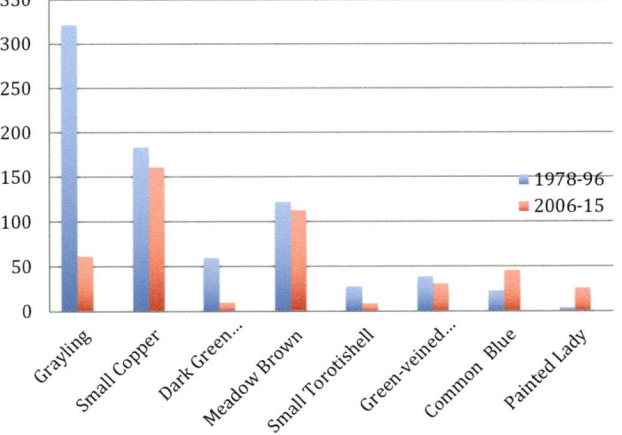

Figure 9.25 Examples of recent changes at Tentsmuir in the populations of some selected butterfly species recorded during monitoring for the UK Butterfly Monitoring Scheme (Fyfe 2017).

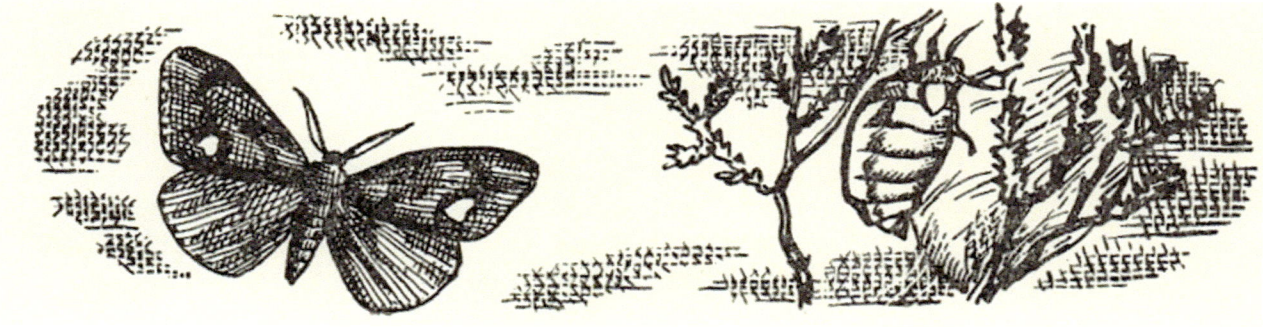

Figure 9.26 The Vapourer Moth (*Orgyia antiqua*) – left male and right the wingless female. (Reproduced with permission from Fullerton (1956). This used to be a common species at Tentsmuir but has not been recorded since the end of the Millennium.

Figure 9.27 Female Emperor Moth (*Saturnia pavonia*) Reproduced with permision from drawing by Len Fullerton.

Figure 9.28 Sand Dart (*Agrotis ripae*). Reproduced with permission (Bryant 2017).

Over the past half century a number of outstanding changes have taken place among the species of Macro Moths that used to frequent the dunes and heaths. Many of the species of Macro-moths that Len Fullerton would have known at Tentsmuir are now gone.

The variety of species that Fullerton observed in the 1950s, he attributed in great part to the nature of the flora. Seventy years ago Tentsmuir had extensive coverage in coastal heathland which was dominated at that time by heather (*Calluna vulgaris*) in the drier regions and Cross-leaved Heath (*Erica tetralix*) in the wetter slack areas. A species that he particularly noted was the Emperor Moth (*Saturnia pavonia*- Figure 9.27) with its handsome caterpillars with green and and carmine pink patches which matched exactly the colour of the heather upon which they fed.

Another species that Fullerton described as common, was the Vapourer Moth (Figure 9.26). This species was a particular curiosity in that the female rarely moves from the cocoon from which she has emerged. Like the Emperor Moth, she also throws out a scent to attract the male, and lays her eggs on the cocoon from which she has emerged. Sadly, this phenomenon can no longer be observed at Tentsmuir as this species has not been recorded there since 2000 (Bryant and Gerald 2017).

Other Large Moth species now similarly absent from Tentsmuir since 2000 AD include, Royal Mantle, Dark Spinach, Broom Tip, V-Moth, Wood Tiger, Sword Grass, Fen Square-spot and Lunar Yellow Underwing (Bryant and Gerald 2017). This change in the moth fauna appears to be due to the changes in the nature of the Tentsmuir flora in recent years. The erosion in the northern part of the Peninsula, where the heather used to be predominant has recently been very extensive. This seems to be a continuing process that has possibly been aggravated by rising sea levels and is likely to continue (see Chapter 3).

Although the moorland area has been greatly reduced in recent years Tentsmuir still survives with an extensive presence of flowering plants such as Orache, Sea Purslane and Sea Rocket which supports a significant Macro-moth fauna. An example of such a coastal species is the Sand Dart (*Agrotis ripae*) which is exclusively a maritime species, occupying the coastal edges of sand-dunes (Figure 9.28) The adults are on the wing in June and July. The caterpillars are nocturnal and feed on the broad leaved vegetation of the coastal fringe during the day but burrow into the sand at night where they also over-winter (Figure 8.26).

Similarly in more landward grasslands with a sufficient presence of nectar-producing flowering plants the

Figure 29 Cinnabar Moth (*Tyria jacobaeae*) (Photo reproduced with permission from Bryant 2017).

Figure 9.30 Six-spot Burnet Moth (*Zygaena filipendulae*) (Photo reproduced with permission from Bryant 2017).

Cinnabar and Six-spot and Burnet moths are frequent examples of day-time flying moths (Figure 9.29-30).

The development of Pine Forest where there was once heathland, has also been a factor in the decline of heathland moths. Nevertheless, despite the loss of some of the heathland moth species at Tentsmuir it is still a rich site for the larger moths. A recent detailed study has found that the number of Macro-moth

Tentsmuir's mammals, butterflies and moths

Rodentia

Rodents (Latin verb *rodere* – to gnaw) make up the largest group of mammals, representing approximately 43 percent of all mammalian species. The Rodentia found at Tentsmuir include, rats, mice, shrews, voles, squirrels, and now also beavers. (Rabbits and badgers are not rodents). Rodents are characterized by having one pair of upper incisors (the chisel-shaped teeth at the front of the mouth), and one pair of lower incisors. These teeth grow continually throughout their life.

The smaller rodents, mice, voles, shrews and moles are an important part of the ecosystem ast they provide the high-protein diet that is needed for the raptors and other predators, such as foxes.

Squirrels

Red Squirrels are nowadays possibly considered to be one of the most engaging and noteable of all the woodland animals to be seen in areas such as Tentsmuir. Scotland is now the Red Squirrel's last major British refuge providing a home for 75 percent of the British Red Squirrel's population which is estimated to total about 140,000. Tentsmuir can be considered fortunate to have these animals and many people enjoy watching them coming to the feeding station where they are a popular public attraction.

White-tailed form of the Red Squirrel

More than one form of Red Squirrel can be seen at Tentsmuir. Particularly striking is the long-established sub-species, long common in Scotland, in which there is a bleaching of the coat, and in particular the tail,

Figure 9.31 Native Scottish form of Red Squirrel (*Sciurus vulgaris leucourus* - Photo Author).

species recorded at present to be 320, which comprises more than half the total number of species known to Scotland as a whole. Consequently, Tentsmuir has to be considered as one of the most important sites for Macro-moths in Scotland (Bryant and Gerald 2017).

which becomes noticeably whiter in early spring. No mention was made of this characteristic until the Welsh naturalist, Thomas Pennant (1726-1798), described it in his *History of Quadrupeds* (1781). This white-tailed variety (*Sciurus vulgaris leucourus*) appears to be unique to Britain and Ireland and is particularly common in Scotland (Figure 9.31).

Harvie-Brown (1844-1916) was the first person to comment in detail on the populations of White-tailed Squirrels as seen at Tentsmuir (see *The Squirrel in Great Britain* - 1880-81). He was a frequent visitor to this part of Fife. At first he thought the whole tail to be merely an accidental bleaching. It was not until 1896 however, that he realized it to be a distinctive racial characteristic which is particularly common among Scottish Red Squirrels.

At various times during the 18th and 19th centuries the Scottish Red Squirrrels have been near to extinction. This was a time when there was a significant reduction in pine forest leading to the 18th century extinction of the Capercaillie. However, this huge bird was re-established in 1837 when the Marquis of Dunblane brought 50 birds from Sweden. Although it is now confined mainly to central and north-east Scotland the occasional Capercaillie has been seen at Tentsmuir (observation of author -1966, there is **as** unconfirmed report of one shot at Tentsmuir in the 1970s).

Whether or not the squirrel decline was generally regretted in forestry circles is open to doubt. Squirrels used to be despised by foresters for damaging trees, and even naturalists condemned them for eating birds' eggs. These attitudes tended to expose the Red Squirrel to a policy of general persecution.

In the 1900s a series of very cold winters seriously reduced the Red Squirrel population in Scotland. The double envionmental misfortune of climatic adversity and the reduction of forest cover raised an awareness that they were in danger of extinction. Whatever the

Figure 9.32 Example of the German population of Black Squirrels that are among the many varieties of squirrel colouration still commonly found in North Germany (Photo Dr. H.Koerner).

dark-coloured squirrels were introduced to Tentsmuir at this time and as a result of inter-breeding there is now a significant number of red squirrels with black-tails present today at Tentsmuir (Figure 9.33).

Grey Squirrels (Sciurus carol)

The Grey Squirrel is a very familiar animal in the United Kingdom, despite not being native. It was introduced from North America during the late 19th Century and since then has displaced the native Red Squirrel across most of England and Wales, though not Scotland and Ireland. It is not just their greater size that makes Grey Squirrels dangerous to Red Squirrels (it is always possible to control numbers by culling if deemed necessary). It is as carriers of the squirrel-pox virus that makes the presence of Grey Squirrels a danger. Grey Squirrels, with a long history of past infections (before they were introduced to Britain) have developed a resistance to the disease. Unfortunately for Red Squirrels, infection with this virus has a 100% mortality rate which results in most infected animals dying within 15 days of being infected. Some animals are even thought to die within 4–5 days of infection. Fortunately for the Red Squirrels, after a successful trapping programme Grey Squirrels have not been seen at Tentsmuir for 4 years (Cunningham pers.com.).

Figure 9.33 Native Scottish Red Squirrel with black-tail which has probably arisen from hybridization. The introduction of Black Squirrels from Germany in the early 1900s after a series of cold winters had decimated the Scottish population. The tail in these squirrels does not whiten in summer (Photo Author).

reason, attempts were made to counteract the decline by importing squirrels from Germany and Scandinavia into Scotland. The most common colouring of Red Squirrels in Central Europe, where over 23 sub-species have been recognised, is fox-red. However, in northern and eastern Germany a black form (*Scirius vulgaris* subsp. *fuscuater*) is particularly common and exists along with the red form (Figure 9.31). It is therefore probable that

Another disease which has become common in Red Squirrels in recent years is Leprosy. Red Squirrels have increasingly been observed with leprosy-like lesions on their heads and limbs. Genetic studies have shown that this infection in squirrels in England, Ireland, and Scotland, is similar to one that circulated in the human population in Medieval times in the British Isles. Red squirrels can therefore be considered as a reservoir for leprosy in the British Isles (Avanzi, del-Pozo et al. 2016). Overall, the risk to people from squirrel leprosy is negligible, as where Red Squirrels have been known to suffer from leprosy for decades there have so far been no recorded human cases. It should however be remembered that visible signs of leprosy in human beings can take as long as 20 years to appear.

Ticks etc.

Ticks, Fleas, and sucking lice, are frequently recorded on Scottish Squirrels. Given the presence of Lyme Disease at Tentsmuir, care should therefore always be taken by human visitors to avoid being bitten by ticks.

Beavers

Beavers are back! After an absence of over four centuries, significant numbers of Eurasian Beavers (*Castor fiber*) are once again to be found in Scotland. They are also present within reach of Tentsmuir, having taken up residence along the banks of the river Tay. Beavers started appearing along in the lower reaches of the river Earn from 2001 onwards. Their arrival was at first explained as due to escapes from enclosures. Whether this introduction was an accident or an illegal introduction is open to question. Beaver enthusiasts such as the *Scottish Wild Land Group* have long been campaigning for the reintroduction of Beavers, stressing what they consider to be the benefits of their reintroduction for biodiversity and arguing that these animals should not be trapped. *Castor fiber*, the Eurasian Beaver (Figure 9.34), was hunted to extinction in the past for its pelt and other products which included *Castoreum* the yellowish secretion of

Figure 9.34 Eurasian Beaver (*Castor fiber* - photo Harold Olsen).

the castor sac. Both beaver sexes have a pair of castor sacs and a pair of anal glands located in two cavities under the skin between the pelvis and the base of the tail.

In 1998, an experimental reintroduction, which had considerable public support, was undertaken by Scottish Natural Heritage (SNH) in partnership with the Scottish Wildlife Trust, the Royal Zoological Society of Scotland, and the Scottish Forestry Commission. The original intention was to undertake a time-limited, five-year trial reintroduction of

Eurasian Beavers at Knapdale, Mid-Argyll. However, this trial met with opposition from salmon fishing bodies as well as various land managers, with the result that it was reduced to a small trial with just a few Beavers at Knapdale. However, this also was eventually abandoned bringing to an end this first authorised attempt at Beaver reintroduction in Scotland.

Despite the end of the trial the Beavers remained. Eradicating a population that has escaped, spread, bred and survived, either through accidents or tacit consent, has proved impossible. Partly, this was due to a considerable number of people who kept their knowledge of the presence of Beavers a secret as they considered it wrong to eradicate a species that they considered to be ecologically and aesthetically valuable, as well as being a potentially useful and engaging native animal. As a result estimates of the number of Beavers in Scotland and especially in the Tay continue to grow (SNH records).

On 24 November 2016, the Scottish Government made the landmark announcement that Beavers were to be allowed to remain in Scotland. This was the first occasion that a mammal formerly thought of by many as disruptive and potentially damaging to aquatic environments, was formally reintroduced into Scotland!

Beavers are frequently described as ecosystem engineers, as they are among the few wild animal species which can significantly change the hydrological characteristics and biotic properties of the landscape as a result of their stream-damming activities.

The Eurasian Beaver (*Castor fiber*) and its American counterpart (*C. canadensis*) can impact on the course of ecological succession, species composition and structure of plant communities. However, the nature of their impact differs. The American Beaver builds extensive dams to raise the water table level so that it is above the entrance to the beaver lodge. The European beavers can build dams but tend to burrow into the banks of rivers in order create their lodges.

Beavers have been reported on several occasions at Tentsmuir and unsuccessful attempts were made to trap one at Morton Lochs where it had dammed the outflow from the north loch. It also felled a number of birches of which one was at least was 18 inches in diameter.

The Eurasian and North American Beavers are similar in their ecological requirements, and require water deep enough to cover the entrance to their lodge or burrow. A food cache is often built next to the lodge or burrow. On small, low gradient streams, dams are frequently built to create an impoundment. On large rivers or in lakes, a lodge containing a food cache may also be built. The Beaver is a keystone riparian species in the European landscape and one which can considerably alter streams and rivers by their activities. (Collen & Gibson 2001).

With regard to Tentsmuir, which is low lying and devoid of any significant gradient-assisted drainage, the presence of Beavers building dams would without doubt increase the extent and duration of the existing winter flooding. In past times, this flooding was already of such an extent and a pronounced characteristic of the area, that it earned Tentsmuir the name of *Scheughy*

Dyke. Any Beaver induced enhancement of the existing tendency for high water tables in winter would likely be detrimental to the ecological stability and habitat variation that is essential for the maintenance of the species richness of the area as well as undermining the physical stability of the local river embankments and in particular those of the river Tay.

Badgers

Badgers (*Meles meles*) have long been present at Tentsmuir. This to be expected as Tayside in general and Tentsmuir in particular provides the mixture of woodlands and pasture, together with light sandy soils which facilitate digging. The Badger faces a dichotomy in its environmental needs between what is optimal for digging setts and that which is best suited for searching for food. Sandy soils aid the digging necessary for their setts. However, worms, which are one of the principal items of their diet, are most readily found in richer soils with high organic matter content. As Tentsmuir Forest is established mainly on a sand substratum this provides an attractive areas for the setts. However, it is not optimal for feeding. This however is not a serious defect due to the proximity to landward of pastures with wet soils containing a high organic matter content supporting the desired worm populations

The Scottish Badgers National Survey, 2006 - 2009 confirmed that the highest density of Badger setts occurred in areas where the dominant habitat was arable farmland, deciduous woodland, or intensive grassland.

Fife is also fortunate that its extensive and widespread Badger populations are believed to be free from bovine tuberculosis and from this point of view they are no risk to cattle. Within the area controlled by Scottish Natural Heritage such infection risks do not appear to be a problem. There are therefore no grounds for their removal by the farming community.

At Tentsmuir it is around the area of Morton Lochs and adjacent woodlands that Badgers occur in considerable numbers. Their habit of crepuscular and nocturnal excursions plus an extensive period of winter hibernation has given the Badgers a reputation for being both elusive and and consequently somewhat mysterious. This however only increases our interest when they are observed.

Figure 9.35. Damage to woodland from badger sett digging (Photo author).

Badgers live collectively in family or clan groups and appear to do so with due care and attention to the hygeine of their dwelling. There is usually a latrine area at a little distance from the sett. They also appear to take care of their dead by either entombing them in a non-visited section of the sett or removing them for internment elsewhere. There is even an account of a female Badger removing her deceased mate by pulling with the help of another male, probably her new mate, and burying the recently deceased in a separate hole (Neal, 1969).

Chapter Ten

Saving the Wilderness

Tentsmuir's origins

Tentsmuir is a striking example of a landscape that has changed almost beyond recognition over the past century. Its use as an area for sport can be traced back to the Middle-Ages with a long history as a Royal Hunting Forest (see Chapter 2). One hundred years ago much of the north-east area of the Peninsula was still largely a shooting estate with the surrounding non-wooded land devoted to arable farming and livestock grazing. These activities were managed in a manner that was much less intensive than that practised in modern times.

At the end of the 1914-18 war Britain had been depleted of much of its useful timber. The creation of the Forestry Commission in 1919, was welcomed as a means of trying to restore the national supply of timber. With what might be considered commendable efficiency at that time, the newly-created Forestry Commission purchased and fenced off a large part of Tentsmuir, almost up to the foreshore (see Chapter 6). Had it not been for the foresight of Dr John Berry, then recently appointed as Director of the Nature Conservancy for Scotland, there might never have been any nature reserve at Tentsmuir. As head of the Nature Conservancy in Scotland at that time he took the initiative to purchase in 1954 as a nature reserve a thin coastal strip of land from the Forestry Commission's holding at Tentsmuir for £120, in the belief that it would advance seawards by natural coastal accretion. Happily, this prediction proved to be correct, and resulted in the reserve that we have to-day (see also Chapter 6).

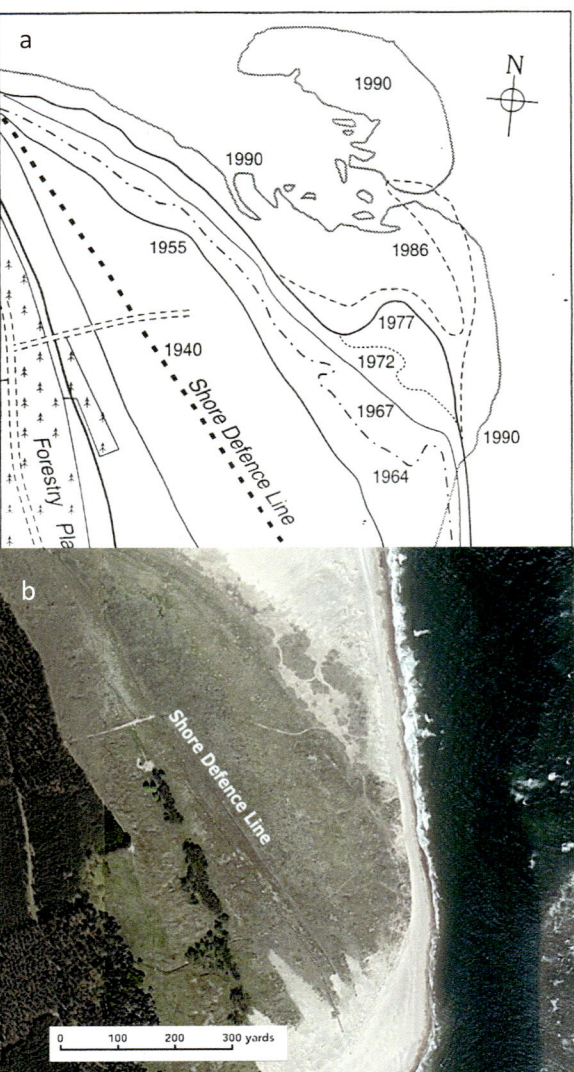

Figure 10.1 Maps of coastal accretion at Tentsmuir: (a) accretion 1856-1990 (Crawford, 1996): (b) erosion as recorded by satellite image (Google Earth 2015).

The land that now constitutes the reserve was therefore a terrain that arose out of the wild – a creation of the winds, waves and sands of the North Sea. The newly-formed dunes and slacks had never belonged to any owner previously, and could therefore be considered as a substantial natural gift of *a pristine wilderness*.

The very small portion that was rescued for a nature reserve, thanks to Dr Berry's initiative, continued to grow. As Dr Berry would have been aware from his boyhood familiarity with the area from probably about 1912, Tentsmuir had already begun to show signs of accretion in the more northern part of the peninsula where there were signs of a seaward advance of the dune systems.

By good fortune this proved true and the rate of accretion accelerated between 1940 and 1990 at an unprecedented rate (Figure 10.1).

Subsequently, the question arose as to whether the reserve should be managed, or left to the forces of nature to mould the landscape, unhindered by human intervention and leave the area to develop as a true wilderness. Such a policy would allow natural forces, both creative and destructive, free to find their own equilibrium and determine the position of the coastline.

Under this policy, there have been times when the biodiversity of the Reserve has flourished, as well other times when the terrain and its flora have suffered grave losses, most commonly due to human intervention, rather than the vicissitudes of nature (see below – on use of goats to counteract an invasion of birch).

Creating the Wilderness Reserve

Current thinking on the creation of nature reserves, usually defines them with a sense of purpose. Consequently, at the time when any particular reserve is created, there is normally a management plan to optimise some aspect of the environment or wild life that appears to be in danger, either from destruction by natural forces or else by human disturbance.

In the past, climate change was not a significant cause of environmental instability or ecological change in Scotland. There have been sudden storms which have caused changes to the landscape, as in 1694 when the unexpected movement of the Culbin Sands during a sandstorm destroyed several farms resulting in the abandonment of a considerable area of land. As a result the Scottish Parliament passed an Act in 1695 banning the removal of dune grasses *for hereafter* in order to prevent similar erosion happening in other places. Such events however, were mainly local and were not repeated, and had no lasting effect on the coastline, providing appropriate remedial measures were taken.

Now however, climate change is causing sea levels to rise, which will inevitably bring about significant alterations in the position of the coastline. Already, such changes are visible at Tentsmuir, where coastal erosion has completely removed the recently accreted portion in the northern part of the reserve (see Figures 10.1-2). At present, there is no great overall loss of terrain, as erosion in the northern part of the reserve has so far been accompanied by extensive deposition of sand to the south (Figure 10.2).

Where before, families could reach the beach from the car park with a walk of just a few hundred yards, parents now have a long weary trudge of up to half a mile or more in order to take their children to the edge of the sea.

The causes of these recent changes are possibly due to alterations in the direction of winds and movement of tidal currents. However, it is equally likely that it is also connected with rising sea levels and therefore linked to the wilds of nature that we are powerless to modify, even although we may have triggered the root cause through fossil fuel burning. The Tentsmuir shoreline and its immediate hinterland is clearly not under human control and can therefore still be regarded as a wilderness zone.

Ecological history of Tentsmuir Point

The creator of the Tentsmuir Nature Reserve, Dr John Berry was born in 1907 into a very different world from today. His mother died when he was very young and he was brought up by devoted aunts. As a result, he roamed the wilds of Tentsmuir accompanying his father on shooting and natural-history trips.

In later years Dr Berry recalled his childhood memories of Tentsmuir as a coastal moorland with much blown sand and many mobile dunes, creating an overall appearance of a desert wasteland. A number of these high dunes had probably formed over buried forest trees. He particularly remembered the region, as having very many high dunes, with some over 50 feet high. There were also areas of extensive barren moorland interspersed with remnants of forest. During periods of high tides, salt water penetrated well into the moor where there was a large brackish pool.

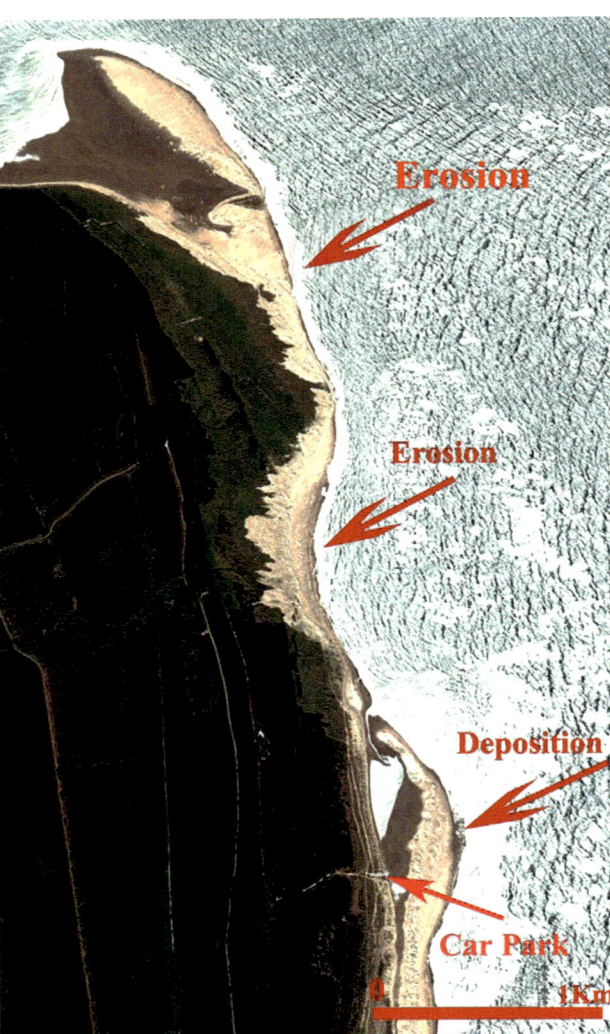

Figure 10.2 View of the Tentsmuir shoreline as recorded by Google Earth 2018.

The influence of high salt concentrations, drought, erosion, and flooding, provided a shifting and unstable landscape with much bare ground, generally evoking an impression of great physical fragility.

Figure 10.3 General view of Tentsmuir Nature Reserve in 1956. View from southeast showing the tree-free slacks and grey dunes. The only significant trees are those that form the flood-line-alder association from alder seeds washed up at the edges of the winter-flooded slacks. The pines in the adjacent forestry plantation have not yet spread onto the Reserve at this time, probably due to the presence of a large rabbit population. Note the straight lines of coastal defence concrete blocks (top right) still at the edge of the shore where they were placed in 1940. (Photo Dr J.K.S. St Joseph).

Figure 10.4 Tentsmuir Nature Reserve in 1972. View from south over approximately the same region as shown in Figure 10.3-. Note the pine trees have now advanced onto the dunes probably as a consequence of the decimation of the rabbit population due to the spread of the Myxoma virus which was illegally introduced to France in 1952 and then spread to Britain the following year where it has long remained as a threat to wild and domestic rabbits. (Photo Dr J.K.S. St Joseph).

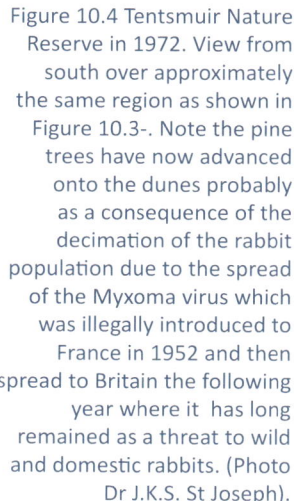

This scenario had a strong effect on the youthful imagination of the young John Berry as he explored the moor, while his father and friends pursued their shooting. A feeling for this semi-barren land lead him to name, along with his father, specific parts of this desolate terrain using Old Testament names. Dr Berry later related that they had to use these secret family names for security measures, as they had discovered that when his father's friends sent telegrams, regarding their visits and the species of birds that had been seen nesting, the telegrams were being intercepted, and egg thieves plundered the eggs from the nests. These information leaks were therefore countered by substituting biblical names for the real names of the locations.

One such biblical name was the *Wilderness of Sin* a coastal area, which in those days was near the Ice House. However, it was in the Shanwell area where most of the rare birds were to be found and where the biblical names were mostly used. The main entry to the Muir, was made from crossing the Lundin Burn, referred to by the Berry family as '*the Jordan.*' The Powie Burn was called the *Brook of Kedron*, after the brook near Jerusalem that David crossed on his flight from Absalom and after which, '*he went up by the ascent of the Mount of Olives, weeping as he went up*' (2 Samuel – Testimony of Gethsemane - Berry pers.com).

The Powie and the Lundin Burns have long been used as boundary markers for fishing rights such as

those granted in 1539 by James V to his Treasurer James Kircaldy. We have therefore in this description of John Berry's youthful impressions of Tentsmuir at the beginning of the 20th century, an image of how it would probably have appeared over the preceding centuries. As its ancient name *Tentis Muiris* indicates, it was a long-accustomed location for transhumance with rough summer grazings. It had also long been used as a place for sport, whether from individual shootings as in more modern times, or as a Royal Hunting Forest for Medieval Kings (see Chapter 2).

For the first half-century since the planting of the forest, its proximity to landward of the reserve had no noticeable effect on the dunes and slacks of the coastal region. Possibly this was due to the high rabbit population that inhabited the dunes and prevented the establishment of pine seedlings (Figure 10.3).

However, by the 1970s the rabbit population had been decimated by the spread of the myxomatosis disease and the pines then advanced onto the heathland sand dunes (Figures 10.4).

Dune invasion by Scots Pine

By the 1970s the presence of this actively-growing and seed-producing pine forest to landward, began to cover much of the area with a dense population of trees (Figure 10.4). The only exceptions were those parts nearest to the sea, and those places where the dune slacks were still regularly inundated in winter.

It became a laborious management policy of SNH in the early 1970s to fell and remove the timber to the foreshore, in the hope of reducing coastal erosion. Sadly, after the clearance of the pines, a treeless landscape did not endure as the pines were rapidly replaced by a dense colonization of birch.

Biological birch control

The changing ecological conditions of the Reserve between 1956 and 1978 as as a result of the invasions of pine, and subsequently birch, were clearly to the detriment of the coastal flora and bird-life for which the Reserve had been created. A desperate situation, obviously needed desperate measures, before the whole of the coastal flora was annihilated. Consequently, it was suggested that some form of biolgcal control by grazing should be tried.

Two young zoology researchers at the University of St Andrews, who were interested in the grazing habits of goats under feral conditions, suggested grazing by goats as a method for controlling the spread of birch. Goat grazing, under a completely feral regime in an extensive area might have been suitable. However, at Tentsmuir it was considered necessary to keep the goats in pens. The resulting intense localized grazing that this caused, had a severe and lasting effect on all the vegetation, even although the goats were moved as thought necessary from one area to another (Figure 10.5). Clearly, from an ecological point of view *the remedy was worse than the disease!*

Imagine therefore, the author's horror on a Christmas morning walk, when he found the goats penned on the most floristically rich area of lichen-rich dune-heath at the north end of the Alder flood-line formation. For years, this area had been pointed out to visiting groups

Saving the Wilderness

Figure 10.5 A remedy that was *worse than the disease*. The disastrous introduction of goats as an aid to conservation at Tentsmuir in 1987 (Photo author).

Onset of coastal erosion

For the first 50 years of its existence the northern portion of the Tentsmuir advanced steadily seawards greatly increasing the area of the reserve. Around 1990 however, there was a marked change, where beginning with the front line of dunes, severe and sustained coastal erosion began. The newly accreted foredunes soon disappeared beneath the waves, and it was not long before the *flood-line alder association*, which had for so long been the the floristically richest and most admired part of the reserve, as noted above by Len Fullerton, began to erode (Figures 10.6-7). In a very short time the alders completely disappeared to be follwed by further erosion.

by Tentsmuir's foremost naturalist, Len Fullerton, as the most floristically-rich site in the whole of Tentsmuir. The goats had now grazed this to the ground, together with the slow growing lichens which have never recovered. After this ecological disaster a management meeting and a conference were eventually called and the severity of the damage high-lighted in a paper entitled ' Tentsmuir, a *National Nature Reserve in Decline*' (Crawford 1996). As a result, the goats were then removed and replaced with cattle. Nevertheless, the reserve has still not recovered the richness of the flora that it had in the first half of the 20th century.

Sadly coastal erosion continues to be the most marked physical feature of the northern section of the Tentsmuir shore. This loss of terrain is however being compensated for by extensive accretion to the south in the Kinshaldy area (Figure 10.2) which is the location

Figure 10.6 The beginning of the end for the Flood-line Alder association (October 2006) once the site of the greatest concentration of flowering plants at Tentsmuir.

that is served by the car park and receives the greatest numbers of visitors, currently estimated to be in the region 130,000 annually.

The extensive erosion to the north of Tentsmuir Point and the Great Slack has removed many of the habitats that were noted in the past for the richness of their flora (see Chapter 2).

Fortunately however, there are already some signs of recovery with the recent recording of the Twin Flower (*Linnaea borealis*) and the Field Gentian (*Gentiana campestris*) (Figures 10.8-9). The most surprising aspect of the return of these species is that the return of the Twin Flower to the pine forest has taken place in the region the Ice House where it was most probably seen by Len Fullerton and Ellis Crapper in the first half of the 20th century.

Figure 10.7 Erosion of the Tentsmuir dunes in 2012. Here only a decade before there had been an extensive area of dunes and slacks (Photo author).

Figure 10.8 A painting of the Twin Flower (*Linnaea borealis*) by Len Fullerton, probably made from a specimen seen at Tentsmuir near the Ice House in the early years of the 20th century (Image in archives of St Andrews University, see also Crapper, 1940).

Figure 10.9 The Field Gentian (*Gentianella campestris*) growing in the Great Slack in 2016 (photo author).

Figure 10.10 Great Slack mostly under water 12th January 2016 (Photo T. Cunningham).

The Great Slack

Attempts were made in the past to remove water from Great Slack by means of a wind pump to aid the growth of trees in the forestry plantation. This did not appear to be effective and it is now considered a better policy from the point of view of conservation diversity to allow the Great Slack to flood in winter. This appears to be beneficial as seen in the recent reappearance there of the Field Gentian (*Gentianella campestris*).

The Great Slack is still prone to flooding (Figure 10.10) and still suffers from tree invasions. Vigilance is constantly necessary to preserve the open heath and dune systems from being invaded by trees. Much effort has been applied to clearing the pines and birches that spread all too readily, particularly onto the dune heaths, as well as the drier slacks. Between 1995 and 2002 a sustained effort was succesful in removing 95 percent of the invading trees and made it possible to restore a substantial area of dune heath (22.2 hectares). Grazing by cattle in summer helps to prevent the trees from spreading without causing the type of damage to the flora which took place when goats were used.

The floristic success of this activity can be seen in the area just to the to the north of the Powie Burn where there is now a successful re-establishment of a vigorous heathland (Figure 10.11) dominated by the CrossLeaved Heath (*Ercica tetralix*). This is the heath species best able to withstand flooding, which is a necessary adaptation as this area has a tendency to flood in wetter winters, despite the proximity, or possibly because of, the proximity of the ancient Powie Burn which readily becomes blocked with sand.

To seawards of the wet heathland is the coastal *Great Slack* which still floods regularly in winter (Figure

10.11). It requires attention however to keep it in a treeles conditions. This is the site where the rare Field Gentian has recently been rediscovered after an absence of some years (Figure 10.9).

In the past this region and the landward facing slopes of the fore-dunes were very rich floristically, with flowering plants such as the Red Centaury flourishing amongst a carpet of Creeping Willow (*Salix repens*) There are now welcome signs that this area is recovering in terms of its range of flowering plants as has been seen in the discovery of the Field Gentian.

A common sight at the edge of the Great Slack in spring used to be the profuse flowering Northern Marsh Orchid (Figure 10.12). Happily the population of these plants and others, that suffered very badly from goat grazing are now showing signs of recovery.

Erosion at the Eden Estuary

It is not just at its northern end (Figures 10.6-7) that the Tentsmuir Peninsula faces dangers from coastal erosion and destruction of valuable wild-life habitats. Similar damage has been taking place at its southern extremity near the Eden Estuary. Here once again, is an example where the short-term misjudged views of a small group of people insisting on a particular form of land management turned out to have disastrous consequences for the preservation of an entire site and its wildlife.

The problem centred on how best to manage the Eden Estuary in relation to the need for stable coastal vegetation and still allow access of the wading birds

Figure 10.11 Restored dune heath dominated by the Cross Leaved Heath, the most flood tolerant of the native Erica species. This area can flood in wet winters. The south bank of the Powie Burn crosses the centre of this figure (Photo author).

Figure 10.12 The Northern Marsh Orchid (*Dactylorhiza purpurella*) growing at the edge of the Great Slack in the 1960s. (Photo author).

to the mud-flats for feeding. In the early 1900s the salt marsh vegetation of the Eden Estuary was sufficiently extensive and stable, for the botany students at St

Andrews University to cross the Estuary on foot from its south side, to Tentsmuir on the north. They were guided in this adventure by their lecturer, Dr Robb, who chose a route that maximized the use of the salt marsh, which at this time must have provided a sufficiently stable substrate to permit this crossing. In more recent times, such a route has never been contemplated by any of his successors, due to extensive erosion of the salt marsh vegetation on both the north and south sides of the estuary.

Once again, the problem of deciding just what should be conserved presented a situation upon which not all were agreed. Frequently, the decisions on how best to preserve an ecological heritage are unfortunately influenced by the personal views of the most vociferous.

The human appreciation of the relative ecological value of bare mud-flats as opposed to salt marshes was unfortunately influenced, at least partly, by whether bird watching was favoured in preference to preserving the flowering plants of the salt marshes and appreciating their vital role in developing a coastal vegetation that was capable of defending the coastline against erosion.

Bird lovers tend to favour mud-flats proximal to the shore for ease of viewing the feeding waders. They therefore fail to appreciate the vital role of coastal vegetation in preventing erosion. Unfortunately, some of those responsible in the recent past for managing the Eden Estuary did not take into account, that preserving the vegetation of the salt marshes need not necessarily preclude the presence of birds.

Figure 10.13 Cord Grass (*Spartina anglica*) in the estuary of the river Eden where it was planted in 1948. The colony was successfully raising the level of the shore and would have provided protection against rising sea levels had it not been systematically eradicated during the 1990s in the mistaken belief that it interfered with the access of wading birds to the mud flats (Photo author).

Figure 10.14 A pilot's view of the Eden Estuary in 2012 showing the extensive area of mudflats that become accessible to wading birds as the tide level begins to fall. (Photo Gordon Pickthall).

An attempt to remedy this situation in favour of salt marsh vegetation was made however in 1948, when the then Professor of Botany at St Andrews University (Professor Graham) surreptitiously planted the hybrid Cord Grass (*Spartina x townsendii; S. maritima x S. alterniflora*) in order to reduce the loss of coastal vegetation.

The hybrid Cord Grass had a reputation particularly in some more southern estuaries e.g. Southampton Water, for aggressive growth to the exclusion of other salt marsh plants. Such a robust and competitive species suggested itself to Professor Graham as possessing the properties needed to resist erosion in the Eden Estuary.

Professor Graham's first planting was washed away. However, he was clearly not a man to give up, and a second planting was made with two dozen more plants. These did not wash away, and after a slow lag period for establishment, succeeded in creating a viable colony which provided a significant vegetation barrier that was successful in beginning to raise the shore-level, thus preventing erosion particularly on the south embankment (Figure 10.13).

If there had been any warning of the possibility of climatic warming and rising sea levels at this time, Professor Graham's single-handed efforts in finding a natural means of defending the coastline would have been greeted with enthusiastic approval.

However, when the area became a local nature reserve it was considered that the Cord Grass might reduce the extent of mud-flats available for wading birds. When the Eden Estuary is viewed from the air (Figure 10.14) it is difficult to understand why the ornithological lobby should consider that there was

Figure 10.15 Aerial view taken on the 13th August 2008 from a helicopter flown from RAF Leuchars. The view looks east over the Eden Estuary at low tide. By this date the Salt Cord Grass has been removed and has exposed a vast extent of bare mud flats. A red arrow shows the subsequent location of a breach of sea wall on the 1st April 2010 after a high storm tide (see Figure 10.16). A century ago much of the area was occupied by a salt marsh. Due to the falling shore level and removal of the Cord Grass this has now been largely replaced by mud flats. Although this might be an advantage to wading birds it has eliminated an essential part of the natural defence of the coastline. (Photo: RAF Leuchars and Professor D.M Paterson).

any shortage of mudflats in the Eden Estuary as a whole. The reason might have been that the south shore was the more accessible for the bird watchers to view the mud-flats as the RAF at Leuchars controlled access to the north shore. Whatever the reason, the management committee for the Eden Estuary decided to actively remove the Cord Grass. This they did with the illegal use of weedkiller below the high tide level.

Sadly, this removal of the Cord Grass served only to aggravate the erosion and lead to extensive reliance on hard defences such as groins and walls which are inherently unstable and in constant need of repair and renewal. The result has been that for over 40 years the area around the northern extremities of the golf courses and the Eden Estuary has been dominated by eroding mudflats and falling shore levels (Figures 10.17-18), threatening the St Andrews golf courses and local farm land with further erosion (Figures 10.15-16).

Ecological restoration

Given the past dependence on unstable artificial coastal defences in the Eden Estuary a new initiative using Sea Club Rush (*Bolboschoenus maritimus*) pioneered by Dr C Maynard (St Andrews University) was greatly welcomed and is now proving successful in stabilizing the shore line of the River Eden (Figure 10.19).

This is a robust species capable of resisting erosion and unlikely to become invasive, which was a fear that has

Figure 10.16 The breach of the sea wall and railway embankment that took place on the south side of the Eden Estuary on the 1st April 2010 (Photo author). The sea wall was originally constructed by French prisoners of war in 1815 and used by Sir Thomas Boutch as a cost cutting measure for the construction of a railway line from St Andrews to Guardbridge as part of his ill-fated Tay Bridge construction that collapsed on the 28th December 1879. The unsatisfactory nature of Sir Thomas Boutch's cost-cutting measures were already apparent after an accident in 1864 on the railway line to St Andrews which was commented on at that time by the Secretary to the Railway Department of the Board of Trade before the Tay Bridge Disaster had taken place in 1879 (Duck 2015).

Figure 10.17 (a) Artificial shore defence wall without any supporting vegetation (Photo. Author). (b) Detail of same wall with no vegetation and which needs constant attention due to falling shore levels (Photo. C. Maynard).

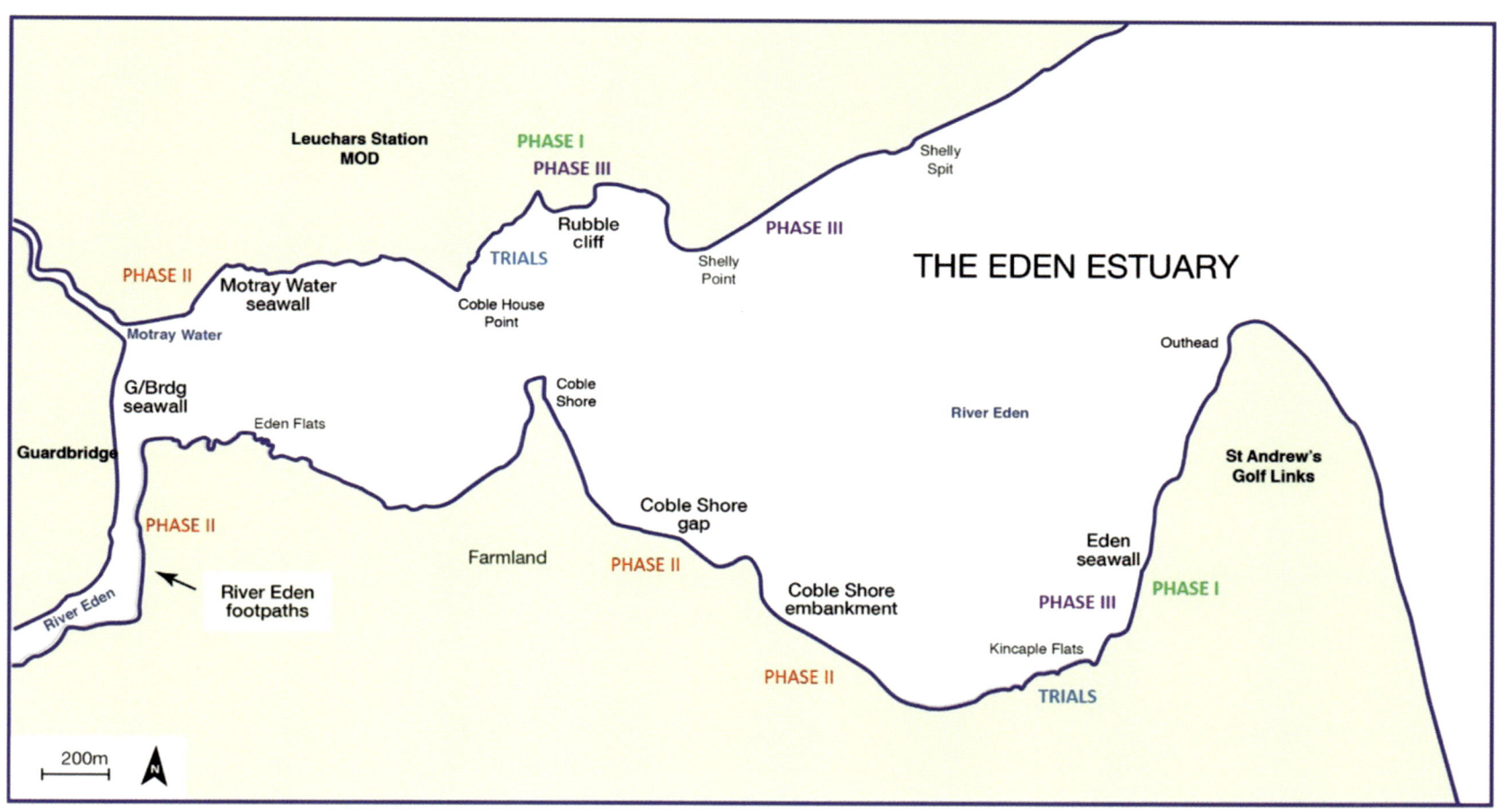

Figure 10.18 Phased plan of ecological restoration of trial plantings of Sea Club Rush (*Boloboschoenus maritima*) to improve sedimentation retention in the Eden Estuary (Maynard, McManus et al. 2011).

been expressed in relation to Cord Grass (see above). When the Eden Estuary is viewed from the air it is difficult to understand in the first place why there was such fear expressed by the ornithological lobby that there might be a shortage of mudflats for the feeding of wading birds in the Eden Estuary (see Figure 10.14).

The use now of Sea Club Rush (see Figure 10.19) should allay these fears. The Club Rush is already proving to be a highly successful species in defending the estuary shore from erosion and provides and secures a robust salt marsh as well as successful defence against falling shore levels (Maynard, McManus et al. 2010).

Conservation and education

Conservation and education are two topics that are inevitably linked. Preserving our natural heritage depends on successive generations having an understanding of the ecology of the world in which we live. In particular, an awareness of habitat vulnerability and its consequences for for the the survival of plant and animal communities and their role in preserving landscapes is essential in planning conservation measures.

The Tentsmuir National Nature Reserve is therefore fortunate that successive reserve managers, as well as managing the Reserve, have devoted considerable time and imagination in capturing the interest not just of the general public, but also in inspiring our younger visitors of the need for proper care for wildlife in the modern world. As has been shown at Tentsmuir in recent years, caring for and understanding the vulnerabilities of our natural surroundings cannot begin too soon in the educational curriculum.

2011

2014

2017

Figure 10.19 Stages over a period of 6 years in the restoration of vegetative cover in the salt marsh on the south bank of the Eden Estuary with Sea Club Rush (Photos courtesy of Dr. C. Maynard).

Figure 10.20 Youthful apprentice surveyors studying sand dune migration (Photo T Cunningham).

Figure 10.21 A group of some of Tentsmuir's youngest visitors being encouraged to take a close look in detail where conservation begins with the soil. (Photo T. Cunningham).

In recent years under the leadership of the Reserve Manager (Tom Cunningham) educational visits for both young and old have been greatly developed (Figures 10.20-21).

Environmental recording

An educated appreciation of the environment, for both young and old cannot be achieved without historical records. Whether or not an area designated as a nature reserve is proving successful, has to be judged against its historical wild-life record. Like any organisation, a nature reserve has to keep its historical records in a manner that is accessible to all. The three decades of Tom Cunningham's responsibility for Tentsmuir have been remarkable in this respect. Thirty years ago the reserve bibliography, before Tom Cunningham took over as manager at Tentsmuir, amounted to nine pages and 300 odd entries. Now it has 102 pages with some 2,010 entries.

To make even casual visitors aware of the Tentsmuir's long history since the end of the ice age an historical *Time Line* of Tentsmuir has been set up covering the past 27,000 years. This trail stretches for over 9 km through the forest with sculptures depicting notable events and developments over the past 27 millennia in a manner that can be appreciated by both young and old.

Hopefully, this vivid illustration of Tentsmuir's history will create an awareness of the interactions between the environment and wildlife and the need not just to preserve, but also to create and manage the habitats that are needed to ensure species preservation.

Tentsmuir is fortunate that together on the one peninsula there is not only an extensive nature reserve but it is also surrounded on all sides by a range of habitats that permit free access to a diversity of landscapes from coastal fringes and woods, to moorlands and lochs. This unique situation was well understood by Dr John Berry when he created the nature reserve in 1954. It was needed then, and this continues to be recognised by Scottish Natural Heritage who have constantly been supporting imaginative educational activities. It is therefore a welcome development that an educational pavilion has been now provided (Figure 10.22), specifically intended for the younger visitors. If the *Wilderness* is to be saved it will the youth of today that have to ensure its future.

Tentsmuir in the future

When the public read, hear, and speak about Tentsmuir it is no longer the flowers and the Eider and Shell Ducks in the the Dunes. Instead, it is the Ospreys that are regularly sighted there, or else, it is the Sea Eagles that now nest in its trees within two miles of the centre of Dundee.

Although the moorland is no longer a safe nesting place for birds such as Eider Ducks and Terns, many can still be seen there during migration. Several species of ducks benefit greatly from the presence of the estuaries of the rivers Tay and Eden which is evident as these areas now have international status as over-wintering grounds for a wide variety of migrant birds.

With the recent closure of the Royal Airforce base at Leuchars, after a century-long presence, there are already signs that the use of the Tentsmuir shore is increasing as a migration route and that the numbers of over-wintering birds are also growing.

Bar-tailed Godwits, Goosanders, Red-breasted Merganser and Long-tailed Ducks are all species that can now be seen as regular visitors outside the breeding season at Tentsmuir .

In winter, Tentsmuir Point and the estuaries of the rivers Tay and Eden are used for feeding and roosting for waders and other birds species along with numerous Waterfowl. These include large numbers of, Oystercatchers, Golden Plovers, Lapwings, Dunlin, Curlew, Cormorant, Shelduck, Goldeneye, Grey Plover, Sanderling, Black-tailed Godwit, Red Throated Divers, Whooper Swans, Great Crested Grebes, Little

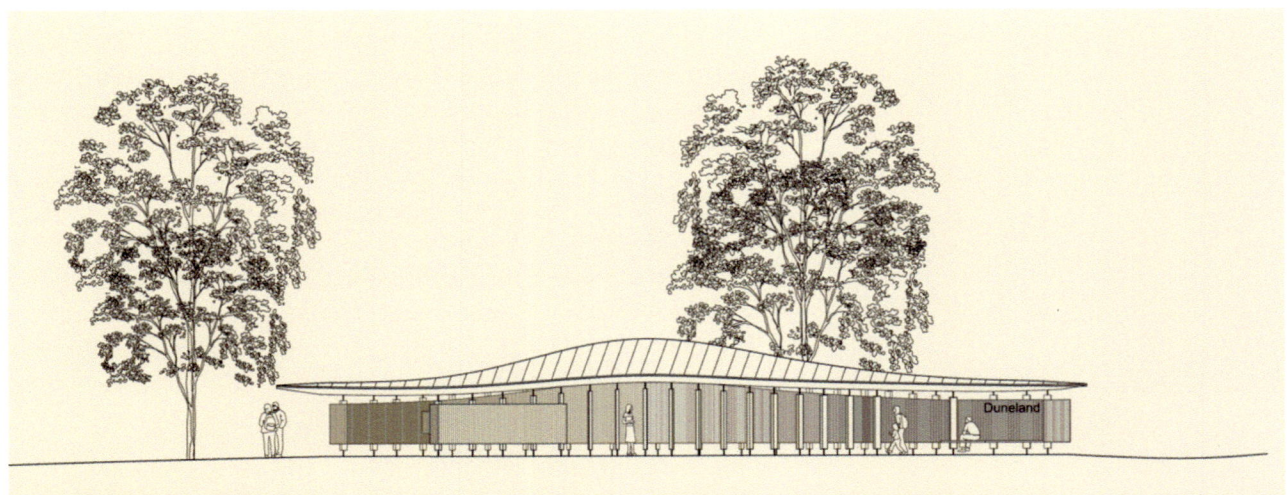

Figure 10.22 North elevation of educational pavilion erected in 2019 and situated near the location of the Icehouse.

Figure 10.23 Painting by Len Fullerton of Bluethroat on Sea Buckthorn (*Hippophae rhamnoides*). A view from the past and a hope for the future. (Reproduced with permission from the daughters of the artist)

Grebes, Mallard, Pochard, Lapwing, Golden Plover, Greenshank and Redshank. The Common Scoter is another winter visitor that can be see on the seen along the shore at Tentsmuir.

The maturing forest that has now been in place for nearly a century has altered the wild life, but at the same time it has provided a range of habitats that previouslydid not exist.

In relation to plants, there are also opportunities to encourage species that help to reduce erosion both on the shore and in the dunes and improve and secure suitable habitats for visiting birds. One such plant is Sea Buckthorn (*Hippophae rhanmoides*) which not only fixes atmospheric nitrogen, thus increasing the fertility of the sandy soil and physically protecting the sand from erosion (Figure 10.23). The bushes grow in thickets that are laden with orange berries in autumn and provide both feeding and shelter for many species of birds.

In recent years climatic warming appears to have resulted in a northward expansion of Sea Buckthorn to such an extent that it is now being regarded as an unwelcome constituent of the flora in Scotland's Nature Reserves. The remedy of total removal of this species by SNH is unfortunate as Sea Buckthorn brings many benefits for the native flora and fauna. It provides highly attractive feeding for birds also. In recent years the ever growing of number visitors to Tentsmuir is welcome, but many are accompanied by dogs which are allowed to run free and inevitably disturb the wild life communities. Thickets of Sea Buckthorn could do much to provide a convenient refuges from such unwanted canine molestation!

A bird that is still observed from time to time on the Sea Buckthorn bushes is the Bluethroat (Figure 10.24). This species was much admired by Tentsmuir's outstanding naturalist and artist Len Fullerton. Unfortunately, there has recently been a wholesale removal of these bushes from the Reserve and along the Fife Coastal Path and elsewhere in order to facilitate access for walkers. It would nevertheless be a welcome move to allow some colonies of Sea Buckthorn thickets to survive

Tentsmuir is not a static environment that should be forced to conform to any human perception of what is natural or normal. Tentsmuir continues to survive by adapting itself to a changing world.

References

Adamson, P. *Early Photographs of St Andrews*. The Archives, University of St Andrews.

Avanzi, C., J. del-Pozo, A. Benjak, K. Stevenson, V.R. Simpson, P. Busso, J. McLuckie, C. Loiseau, C. Lawton, J. Schoening, D.J. Shaw, J. Piton, L. Vera-Cabrera, J.S. Velarde-Felix, F. McDermott, S.V. Gordon, S.T. Cole and A.L. Meredith. 2016. Red squirrels in the British Isles are infected with leprosy bacilli. *Science* 354 (6313): 744-747.

Bagdanaviciute, I., L. Kelpsaite, and T. Soomere 2015. Multi-criteria evaluation approach to coastal vulnerability index development in micro-tidal low-lying areas. *Ocean & Coastal Management* 104: 124-135.

Ballin, T.B. 2016. Rising waters and processes of diversification and unifications in material cutures: the flooding of Doggerland and its effect on north-west European prehistoric populations between 13,000 and 11,500 cal BC. *Journal of Quaternary Science* 1999: 1-11.

Ballin, T.B., and H.B. Bjerck 2016. Lost and found twice: Discussion of an early post-glacial single-edged tanged point from Brodgar on Orkney, Scotland. *Journal of Lithic Studies* 3(1): 31-50.

Bauch, J., and H. Berndt 1973. Variability of chemical composition of pit membranes in bordered pits of gymnosperms. *Wood Science and Technology* 7: 6-19.

Baxter, E.V., and L.J. Rintoul 1953. *The Birds of Scotland*. Oliver and Boyd, Edinburgh.

Berry, J. 1930. Notes on the movements of duck off Tentsmuir Point, Fife. *Scottish Nature* 1930: 43-46.

Berry, J. 1939. *The status and distribution of Wild Geese and Wild Duck in Scotland*. Cambridge University Press, Cambridge.

Berry, W. 1894. On the introduction of grouse to the Tents Muir in Fife. *Annals of Scottish Natural History* 12: 197-203.

Berry, W. 1894. *Hippophae. Journal of Experimental Botany* 6: 303-311.

Bexton, S., D. Thompson, A. Brownlow, J. Barley, R. Milne and C. Bidewell 2012. Unusual Mortality of Pinnipeds in the United Kingdom Associated with Helical (Corkscrew) Injuries of Anthropogenic Origin. *Aquatic Mammals* 38(3): 229-240.

Bocher, P., F. Robin, B. Deceuninck, and E. Caillot 2013. Distribution, phenology and long-term population trends in Black-tailed Godwits *Limosa limosa* wintering in coastal France. *Acta Ornithologica* 48: 141-150.

Bond, G. 1955. An isotopic study of the fixation of nitrogen associated with nodulated plants of *Alnus, Myrica. Journal of Experimental Botany* 6 (17): 303-311.

Bonsall, C., C. Tolan-Smith, and A. Saville 1995. Direct dating of Mesolithic antler and bone artifacts from Great Britain: new results for bevelled tools and red deer antler mattocks. *Mesolithic Miscellany* 16: 2-10.

Boswell, J. 1791 *The Life of Samuel Johnson, LL.D.* (reprint 1900). Swan Sonnenschein & Co. Ltd, London.

Browne, M.A.E., and J. Jarvis 1983. Late Devensian marine erosion in St Andrews Bay, east-central Scotland. *Quaternary Newsletter* 41: 11-17.

Brownlow, A., J. Onoufriou, A. Bishop, N. Davison, and D. Thompson 2016. Corkscrew Seals: Grey Seal (*Halichoerus grypus*) Infanticide and Cannibalism May Indicate the Cause of Spiral Lacerations in Seals. *PLOS ONE* 11(6)

Bryant, D.M. and A.L. Gerald 2017. *The larger moths of Tentsmuir: 1976 to 2017.* Draft report to Scottish Natural Heritage and Forestry Commission.

Camphuysen, C.J., C.M. Berrevoets, H. Cremers, A. Dekinga, R. Dekker, B.J. Ens, T.M. van der Have, R.K.H. Kats, T. Kuiken, M.F. Leopold, J. van der Meer, and T. Piersma 2002. Mass mortality of common eiders (*Somateria mollissima*) in the Dutch Wadden Sea, winter 1999/2000: starvation in a commercially exploited wetland of international importance. *Biological Conservation* 106: 303-317.

Candow, R.D.M. 1989. *Prehistoric Morton: The story of the Mesolithic discoveries at Morton Farm on Tentsmuir in North East Fife.* David Winter and Son, Dundee.

Carter, S. 1997. *The Archaeology of Tentsmuir: A synthesis and interpretation of existing records.* Report Commsioned by Fife Council and Historic Scotland, Edinburgh.

Cieslak, M. 1980. European populations of Osprey *Pandion-haliaetus* present state and threats. *Przeglad Zoologiczny* 24: 123-136.

Coles, B.J. 1998. Doggerland: a speculative survey. *Proceedings of the Prehistoric Society* 64: 45-81.

Coles, B.J. 2000. Doggerland: the cultural dynamics of a shifting coastline. In *Coastal and estuarine environments: sedimentology, geomorphology and geoarchaeology*. K. Pye and J.R.L. Allen, editors. Geological Society Special Publication, Vol. 176: 393-401. London.

Coles, J.M. 1964. Scottish Middle Bronze Age metalwork. *Proceedings of the Society of Antiquaries of Scotland* 97 (1963-64): 82-156.

Coles, J.M. 1971. The early settlement of Scotland: excavations at Morton Fife. *Proceedings of the Prehistoric Society* 38: 284-366.

Coles, J.M. 1983. Morton revisited. In *From the Stone Age to the 'Forty-Five – Studies presented to R.B.K. Stevenson*. A. O'Connor and D.V. Clarke, editors. John Donald, Edinburgh.

Cowing, E. 2012. Archaeologists unearth Stone Age dwelling on the banks the of new Forth crossing. In *The Scotsman*, Edinburgh.

Crapper, E. 1939-1940. *The Flora of Tentsmuir, Fife Parts 3-4*. The Archives, University of St Andrews.

Crawford, B. 2017. The Medieval Leuchars Bowl. *History Scotland* 17 (4): 42-45.

Crawford, R.M.M. 1996. Tentsmuir Point: A national nature reserve in decline? In *Fragile environments: the use and management of Tentsmuir.*

G. Whittington, editor. Scottish Cultural Press, Aberdeen: 65-88.

Crawford, R.M.M. 2008. *Plants at the Margin – ecological limits and climate change*. Cambridge University Press, Cambridge.

Crawford, R.M.M., and D. Wishart 1966. A multivariate analysis of the development of dune slack vegetation in relation to coastal accretion at Tentsmuir, Fife. *Journal of Ecology* 54: 729-743.

Crawford, R.M.M., K. Studer-Ehrensberger, and C. Studer 1997. Flood-induced change on a dune slack observed over 24 years. In *The ecology and conservation of European dunes*. F. Garcia Novo, R.M.M. Crawford, and M.C. Diaz Barradas, editors. University of Seville, Seville.

Cunningham, T.D. 2015-16. *Annual Report Tentsmuir National Nature Reserve*.

Deith, M.R. 1983. Molluscan calendars: the use of growth-line calendars to establish seasonality of shellfish collection at the Mesolithic site of Morton, Fife. *Journal of Archaeological Science* 10: 423-440.

Deith, M.R. 1986. Subsistence strategies at a Mesolithic camp site: evidence from stable isotope analyses of shells. *Journal of Archaeological Science* 13: 61-78.

Dobson, M.J. 1994. Malaria in England: a geographical and historical perspective. *Parassitologia* 36: 35-60.

Doody, P. 2015. Distribution of Sand Dunes in Great Britain. J.P. Doody, editor. UK Joint Nature Conservation Committee. http://www.coastalwiki.org/wiki/Distribution_of_Sand_Dunes_in_Great_Britain

Duck, C. and D. Thompson 2008. UK grey seal pup production at annually monitored breeding colonies in the main island groups. *NAMMCO Scientific Publications* 6: 69-78.

Duck, C.D. and C.D. Morris 2016. *Surveys of harbour and grey seals on the south-east (border to Aberlady Bay) and south-west (Sound of Jura to Solway Firth) coasts of Scotland, in Shetland, in the Moray Firth and in the Firth of Tay in August 2015*. Scottish Natural Heritage Commission Report No. 929.

Duck, E. 2011. *This shrinking land: climate change and Britain's coasts*. Dundee University Press, Dundee.

Duck, R. 2015. *On the Edge: coastlines of Britain*. Edinburgh University Press, Edinburgh.

Elkins, N., J.B. Reid, A.W. Brown, D.G. Robertson, and A.-M.W. Smout 2003. *The Fife Bird Atlas*. Woodlands Studios, Dunfermline.

Elkins, N., J.B. Reid, and A.W. Brown 2016. *The Breeding and Wintering Birds of Fife – an atlas for 2007-2013*. Scottish Ornithologists' Club, Aberlady.

Evans, R.J., J.D. Wilson, A. Amar, A. Douse, A. MacLennan, N. Ratcliffe, and D.P. Whitfield 2009. Growth and demography of a re-introduced population of White-tailed Eagles Haliaeetus albicilla. *Ibis* 151: 244-254.

Fullerton, L. 1956. *Tentsmuir – a changing landscape*. International Union for the Protection of Nature.

Fyfe, G.J. 2017. *A report on butterfly abundance and flight periods at Tentsmuir Nature Reserve from 1978 to 2015*. Scottish Natural Heritage Commissioned Report.

Gifford, J. 1988. *The Buildings of Scotland – Fife*. Penguin Books, London.

Glenn, V. 2003. *Romanesque and Gothic: Decorativer Metalwork and Ivory Carvings in the Museum of Scotland*. National Museums of Scotland, Edinburgh.

Grierson, J. 1962. A check-list of the birds of Tenetsmuir, Fife. *Scottish Birds* 2 Special Supplement: 113-163.

Grierson, J. 1962. A check-list of the birds of Tentsmuir, Fife. *Scottish Birds* 2 Special Supplement: 113-163.

Hammond, M. 2013. The paradox of medieval Scotland, 1093-1286 (Introduction (1-52). In *New Perspectives on Medieval Scotland 1093-1286* (Studies in Celtic History). M. Hammond, editor. Boydell Press, Woodbridge.

Hanson, N., D. Thompson, C. Duck, and J.M. Baxter 2017. Harbour Seal (*Phoca vitulina*) abundance within the Firth of Tay and Eden Estuary, Scotland: recent trends and extrapolation to extinction. *Aquatic Conservation: Marine and Freshwater Ecosystems* 27: 268-281.

Harvie-Brown, J.A. 1875. *Travels of a naturalist in Northern Europe; Norway 1871, Archangel 1872, Petchora, 1875*. T.F. Unwin, London.

Hasenohr, G., and M. Zink 1992. *Dictionnaire des lettres françaises: Le Moyen Age. Collection: La Pochothèque*. Fayard, Paris.

Hultén, E., and M. Fries 1986. *Atlas of North European vascular plants north of the tropic of Cancer*. Koeltz Scientific Books, Königstein.

Ibarrola, I., X. Larretxea, E. Navarro, J.I.P. Iglesias, and M.B. Urrutia 2008. Effects of body-size and season on digestive organ size and the energy balance of cockles fed with a constant diet of phytoplankton. *Journal of Comparative Physiology B - Biochemical, Systemic, and Environmental Physiology* 178: 501-514.

Jenkins, P.A., R.W. Duck, and J.S. Rowan 2005. Fluvial contribution to the sediment budget of the Tay Estuary, Scotland, assessed using mineral magnetic fingerprinting. In *Sediments Budgets 1 – Proceedings of symposium S1 -Seveneth IAHSScientific Assembly*. Vol. 291. IAHS Publication, Foz do Iguacu, Brazil.

Kettle, R.R. 1796. Parish of Leuchars. In *The Statistical Account of Scotland* Vol. 18. S.J. Sinclair, editor.

Little, L.R., and M.A. Maun 1996. The '*Ammophila* problem' revisited: a role for mycorrhizal fungi. *Journal of Ecology* 84: 1-7.

Lockwood, W.B. 1993. *The Oxford Dictionary of British Bird Names*. Oxford University Press, Oxford.

Longworth, I.H., R.D.M. Candow, R. Crerar, and D. Henderson 1966-7. Further discoveries at Brackmont Farm and Tentsmuir, Fife. *Proceedings of the Society of Antiquaries of Scotland* 99: 60-92.

Lovat, Simon Joseph Fraser, 14th Lord Lovat 1911. The grouse in health and in disease. In *Final report of the committee of enquiry on grouse disease* Vol. 2. L.L. Chairman, editor. Published for the committee of enquiry on grouse disease by Smith, Elder and Co., London.

MacGillivray, W. 1837-1852. *A history of British birds, indigenous and migratory*. Scott, Webster and Geary, London.

MacLeod, R.W. 1996. *Lairds and farmers in Fife*. Cupar, Fife.

Maynard, C.E., J. McManus, R.M.M. Crawford and D.M. Paterson 2010. Saltmarsh sedimentation patterns in the Eden Estuary (Scotland): a comparison between natural and transplanted marsh. *Plant Ecology and Diversity* 4(4): 103-113.

McManus, J., and A. Wal 1996. Sediment accumulation mechanisms on the Tentsmuir coast. In *Fragile environments: The use and management of Tentsmuir NNR, Fife*. G. Whittington, editor. Scottish Cultural Press, Edinburgh.

McManus, J., and S.A.K. Alizai. 1987. Variations in marsh surface levels in the upper Tay Estuary. *Proceedings of the Royal Society of Edinburgh* 92B: 345-358.

Melvin, J. 2011. *James Colquhoun Irvine: St Andrews' Second Founder*. John Donald, Edinburgh.

Millais, J.G. 1902. *The Natural History of British Surface-Feeding Ducks*. Longmans, London.

Moore, P.D. 1996. Mystery of moribund marram. *Nature* 380: 286-286.

Newland, D. and R. Still 2010. *Britain's Butterflies*. U.K., Wild Guides Ltd, Princeton University Press, Woodstock.

Nielsen, J., R.B. Hedeholm, M. Simon, and J.F. Steffensen 2014. Distribution and feeding ecology of the (*Somniosus microcephalus*) in Greenland waters. *Polar Biology* 37: 37-46.

Ovington, J.D. 1951. The afforestation of Tentsmuir sands. *Journal of Ecology* 39: 363-375.

Penant, T. 1771. *A Tour of Scotland, 1769*. Chester.

Pennycuick, C.J., T.A.M. Bradbury, O. Einarsson, and M. Owen 1999. Response to weather and light conditions of migrating Whooper Swans *Cygnus cygnus* and flying height profiles, observed with the Argos satellite system. *Ibis* 141: 434-443.

Petty, S.J. 1996. History of the Northern Goshawk *Accipiter gentilis* in Britain. In *The Introduction and Naturalisation of birds* S.J. Holmes and J. Simons, editors. HMSO, London.

Petty, S.J. 2002. Northern Goshawk. In *The Migration Atlas: Movements of Birds of Britain and Ireland*. C.V. Wenham, M.P. Thoms, J.H. Marshant, and et.al., editors. Poyser, London.

Pinno, B.D., S.M. Landhaeusser, P.S. Chow, S.A. Quideau, and M.D. MacKenzie 2014. Nutrient uptake and growth of fireweed (Chamerion angustifolium) on reclamation soils. *Canadian Journal of Forest Research-Revue Canadienne De Recherche Forestiere* 44: 1-7.

Pryor, F. 2004. *Britain BC: Life in Britain and Ireland before the Romans*. Harper Perennial, London.

Ramsdale, C.D., and N. Gunn 2005. History of and prospects for mosquito-borne disease in Britain. *European Mosquito Bulletin*. 20: 15-30. Report lodged with Scottish Natural Heritage.

Ritchie, R.L.G. 1954. *The Normans in Scotland*. Edinburgh University Press, Edinburgh.

Ross, S. 1993. The Culbin Sands – A mystery unravelled. In *Moray: Province and People*. S. W.D.H., editor. The Scottish Society for Northern Studies, Edinburgh.

Saville, A. 2004. *Mesolithic Scotland and its Neighbours*. Society of Antiquaries of Scotland, Edinburgh.

Saville, A. 2008. The beginning of the Later Mesolithic in Scotland. In *Man – Millennia – Environment: Studies in Honour of Romuald Schild*. Z. Sullgostowska and A.J. Tomaszewski, editors. Polish Academy of Sciences, Warsaw.

Seaver, G. 1933. *Edward Wilson of the Antarctic: naturalist and friend*. John Murray, London.

Seaver, G. 1937. *Edward Wilson: nature lover*. John Murray, London.

Seivwright, L.J., S.M. Redpath, F. Mougeot, F. Leckie, and P.J. Hudson 2005. Interactions between intrinsic and extrinsic mechanisms in a cyclic species: testosterone increases parasite infection in red grouse. *Proceedings of the Royal Society B - Biological Sciences* 272: 2299-2304.

Shand, S.J. 1908. Note upon a group of kitchen-middens on Tents Muir near Guardbridge. *Transactions of the Perthshire Society of Natural Science* 4: 187-188.

Simpson, G.G. 1985. The *Familia* of Roger de Quincy, Earl of Winchester and Constable of Scotland. In *Essays on the Nobiity of Medieval Scotland*. K. Stringer, J., editor. John Donald, Edinburgh.

Skieresz-Szewczyk, K., and H. Jackowiak 2016. Morphofunctional study of the tongue in the domestic duck (*Anas platyrhynchos f. domestica*, Anatidae): LM and SEM study. *Zoomorphology* 135: 255-268.

Smart, R.S. 2015. *The St Andrews portion of the protocol book of William Gray 1553-59*.

Smith, D.E., M.H. Davies, C.L. Brooks, T.M. Mighall, S. Dawson, B.R. Rea, J.T. Jordan, and L.K. Holloway 2010. Holocene relative sea levels and related prehistoric activity in the Forth lowland, Scotland, United Kingdom. *Quaternary Science Reviews* 29: 2382-2410.

Smith, D.E., N. Hunt, C.R. Firth, J.T. Jordan, P.T. Fretwell, M. Harman, J. Murdy, J.D. Orford, and N.G. Burnside. 2012. Patterns of Holocene relative sea level change in the North of Britain and Ireland. *Quaternary Science Reviews*. 54: 58-76.

Smith, T.L. 1948. Growth and decline of an artificial grouse moor. *The Scottish Naturalist* 60: 99-106.

Smout, A.-M. 1986. *The Birds of Fife*. John Donald, Edinburgh.

Smout, A.-M. 1996. The Birds of Tentsmuir, 1880-1990; An ecological catastrophe? In *Fragile Environment; The use and management of Tentsmuir NNR, Fife*. G. Whittington, editor. Scottish Cultural Press, Edinburgh.

Smout, A.-M. and P. Kinear 1993. The Butterflies of Fife. *Fife Nature*: 1-28.

Smout, A.-M. and P. Kinnear 1993. *The Butterflies of Fife: A Provisional Atlas*. Scottish Nautral Heritage: Fife Nature Biological Record Centre.

Studer-Ehrensberger, K., C. Studer, and R.M.M. Crawford 1993. Competition at community boundaries: mechanisms of vegetation structure in a dune-slack complex. *Functional Ecology* 7: 156-168.

Sturt, F., D. Garrow, and S. Bradley 2013. New models of North West European Holocene palaeogeography and inundation. *Journal of Archaeological Science* 40: 3963-3976.

Taylor, G.S., A.A. Crowder, and P. Rodden. 1984. Formation and morphology of an iron plaque on the roots of *Typha latifolia* L. grown in solution culture. *American Journal of Botany* 71.

Taylor, M. 2010. *RSPB Britain Birds of Prey*. Bloomsbury Publishing, London.

Taylor, S., with G. Márkus 2010. *The place-names of Fife*; Vols 1-4 Shaun Tyas, Donington.

Vasskog, K., N. Waldmann, S. Bondevik, A. Nesje, E. Chapron, and D. Ariztegui 2013. Evidence for Storegga tsunami run-up at the head of Nordfjord, western Norway. *Journal of Quaternary Science* 28: 391-402.

Wickham-Jones, C., and M. Dalland 1998. A small Mesolithic site at Craighead Golf Course, Fife Ness, Fife. *Tayside and Fife Archaeological Journal* 4: 1-19

Wilson, J.H. 1910. *Nature study rambles round St Andrews*. W.C. Henderson and Son, St Andrews.

Woodall, P. 2001. Family *Alcedinidae* (Kingfishers). In *Handbook of the Birds of the World*. J. del Hoyo, A. Elliott, and J. Sargatal, editors. Lynx Edicions, Barcelona.

Wrånes, E. 1988. Masserdød av aerfugl på Sørlandet vinteren 1981-81. *Våre Fugelfauna* 11: 71-74.

Index

Afforestation, 29, 33, 51, 95, 97, 105

Agricultural improvements, 18

Alder, 34–35, 45, 57–59, 62, 75–76, 80, 167–169

Badgers, 157, 162–163

Beavers, 141, 157, 160–161

Berry, John, 51, 84, 86–87, 92, 96, 113, 164, 166–168, 179

Berry, William 29, 84, 86–92, 94

Birch control, 168

Boswell, James 12

Boundary, 2, 20–21, 35, 42, 66, 73–74, 167

Bronze Age, 9–11, 16

Butterflies, vi, 91, 141, 146–151, 153–154

Churchill, 29–30

Coastal accretion, v, 36, 40, 42, 44, 46, 164

Conservation, v, 32, 63, 69, 84, 86, 102, 110, 118, 131, 169, 171, 178–179

Crapper, Ellias, 54, 56–58, 60–61, 65, 81, 83, 147, 170

Crawford, William, 17

Danish fleet, 13

de Quincy, Robert 14–16

de Quincy, Roger, 18

Doggerland, 6–9

Dr Johnson, 12

Dr John Berry (see also William Berry), 87, 92, 96–98, 102, 114–116

Ducks, 28, 42, 49–50, 86–87, 96–97, 102, 112, 118–123, 128, 131–133, 138, 142–145, 176, 180

    Eider, 86, 97, 112, 132–134, 180

    Gadwall, 102, 121–122

    Shelduck, 86, 90, 96–97, 102, 120, 132–133, 180

Drains, 23, 99–100, 105, 107

Dune slacks, 23, 34–36, 57, 59, 70, 74, 81, 83, 150–151, 168

Dunes, v, 20, 23, 31–37, 39–46, 49–52, 54–56, 59–66, 68–69, 80, 89, 96, 105, 107–108, 120, 130, 132–133, 138, 141, 146–149, 152, 155–156, 165–170, 172, 180, 182

Dykes, 13, 101, 162

Earlshall, 9–11, 18–19, 87, 120, 130

Education, 91, 178

Egypt, 15

Erosion, 37, 39, 41, 43–44, 49, 53, 55, 58–59, 63, 65–66, 76–77, 79, 89, 98, 141, 156, 164–166, 168–170, 172–175, 177, 182

Flooding, 22–23, 28, 31, 35–38, 50, 53, 70–76, 80–82, 89, 93, 101, 161, 166, 171

    fresh water flooding, 75

    salt water flooding, 36, 50–53, 70–80

Fullerton, Len, 57–58, 86, 147, 154–155, 169–170, 181–182

Geese, 96, 111–116

    Barnacle, 112, 115–116

    Bean, 112–113

    Brent, 112, 116

    Canada, 62, 112, 146

    Greylag, 112–115

    Pink-footed, 111–113, 115

    White-fronted, 112–113, 116

General Sikorski, 29–30

German spies, 29

Goats, 62, 165, 168–169, 171–2

Godwits, 117–118, 130, 138, 180

    Bar-tailed, 117–118, 138, 180

    Black-tailed, 117–118, 130, 138, 180

Grouse, 28–29, 50, 86–87, 89–94, 96, 125

History, 1, 3, 12, 15, 18, 25, 42, 44, 46, 48, 84–85, 99, 131, 139, 159, 164, 166, 179

    Medieval, v, 12–16, 18–19, 22, 25, 84, 127, 159, 168

Holocene, 1–2, 5–9, 22

Kingfisher, 128, 130

King Malcolm IV, 14

King Robert I, 17

Lepidoptera, 146–147

Leuchars, 11–19, 22–23, 25–26, 29, 99, 113, 130, 175, 180

Leuchars Castle, 15–19, 99

Longshore drift, 37, 46

Lord Lovat, 90–92

Malaria, 25–26

Mammals, 141–145, 15, 161

Morton, 2–5, 8–9, 11, 22, 42, 51, 86–87, 91, 99–104, 112–113, 117, 119–124, 127–130, 132, 134, 147, 161–162

Moths, vi, 91, 141, 146–147, 154–157

Norman, v, 12–14, 16–17, 20–21

Orabile, 13–15

Osprey, 124–126

Pine invasion, 168

Pleistocene, 1–2, 49

Plovers, 128, 137–138, 180–182

Powie Burn, 19–20, 35, 73, 81, 107–108, 167, 171–172

Prehistory, 1, 8–9

    Palaeolithic, 8

    Mesolithic, 2–9, 22, 36, 42, 49

    Neolithic, v, 9–11

    Bronze Age, v, 9–11, 16

    Iron Age, v, 9–11

Restoration, 18, 71, 102–103, 128, 175, 177–178

Rodentia, 157

Salmon fishing, 19–22, 28, 42, 161

Sand dunes, 20, 32–33, 41–44, 47–49, 54–56, 62–63, 66–69, 107, 168

Slacks, 23, 31, 34–36, 39–40, 42, 44, 51, 54–59, 62–63, 69–71, 73–76, 79–83, 89, 101, 107, 113, 132–133, 146, 150–152, 155, 165, 167–168, 170–171

Salt, 36, 38, 50–51, 53, 70–71, 75–79, 93, 111–112, 133, 166, 172–175, 177–178

Sawbills, 119

    Red-breasted Merganser, 119, 180

    Goosander, 119

Sea-level, 1, 69

Seals, 141–146

    Common seal, 143

    Grey seal, 141–146

Squirrels, 141, 157–160

    Black squirrel, 158–159

    Grey squirrel, 159

    Red squirrel, 141, 157–160

St Andrews, 12–15, 17–19, 37, 39, 45–47, 56–58, 63, 65, 84–85, 112, 118–119, 131–132, 143, 147, 168, 170, 172, 174–176

St Andrews University, 56–58, 143, 147, 170, 172, 174–175

St Athernase Church, 14–15

Tentsmuir, 1–2, 5–6, 9–13, 18–20, 22–26, 28–43, 45–66, 69–76, 79–81, 83–102, 104, 106, 108–128, 130–135, 137–138, 141–144, 146–147, 149–162, 164–170, 172–173, 178–180, 182

War, 27, 29, 33–34, 44, 62, 81, 86–87, 92, 95, 164, 176

Coastal defences, 29, 33, 175

German spies, 29

Polish troops, 30

Wetlands, 24, 31, 36, 38, 70–71, 75, 98–99, 101, 136, 139, 141, 146

White-tailed Eagle, 123–125

Whooper Swans, 101–102, 112, 116–117, 139, 180

Wigeon, 102, 120, 122, 132, 138–139

Wilson, Edward, 91–93, 141